T0344590

Film Silicon Science and Technology

MATERIALS RESEARCH SOCIETY
SYMPOSIUM PROCEEDINGS VOLUME 1536

Film Silicon Science and Technology

Symposium held April 1–5, 2013, San Francisco, California U.S.A.

EDITORS

Paul Stradins

National Renewable Energy Laboratory
Golden, CO, U.S.A

Akram Boukai

University of Michigan
Ann Arbor, MI, U.S.A

Friedhelm Finger

Forschungszentrum Jülich GmbH
Jülich, Germany

Takuya Matsui

National Institute of Advanced Industrial
Science and Technology
Tsukuba, Ibaraki, Japan

Nicolas Wyrsch

Ecole Polytechnique Fédérale de Lausanne
Neuchatel, Switzerland

Materials Research Society
Warrendale, Pennsylvania

CAMBRIDGE
UNIVERSITY PRESS

Shaftesbury Road, Cambridge CB2 8EA, United Kingdom

One Liberty Plaza, 20th Floor, New York, NY 10006, USA

477 Williamstown Road, Port Melbourne, VIC 3207, Australia

314–321, 3rd Floor, Plot 3, Splendor Forum, Jasola District Centre, New Delhi – 110025, India

103 Penang Road, #05–06/07, Visioncrest Commercial, Singapore 238467

Cambridge University Press is part of Cambridge University Press & Assessment, a department of the University of Cambridge.

We share the University's mission to contribute to society through the pursuit of education, learning and research at the highest international levels of excellence.

www.cambridge.org
Information on this title: www.cambridge.org/9781605115139

Materials Research Society
506 Keystone Drive, Warrendale, PA 15086
http://www.mrs.org

First published 2013

CODEN: MRSPDH

A catalogue record for this publication is available from the British Library

ISBN   978-1-605-11513-9   Hardback

CONTENTS

THIN FILM SILICON SOLAR CELLS

*Invited Paper

NOVEL SILICON-BASED DEVICES AND SOLAR CELLS

MATERIALS AND DEVICES CHARACTERIZATION AND SIMULATION

DEFECTS AND TRANSPORT

NANOSTRUCTURED SILICON AND RELATED NOVEL MATERIALS

DEDICATION

The editors dedicate this volume,
Film Silicon Science and Technology,
to J. David Cohen and Stanford Ovshinsky.

MATERIALS RESEARCH SOCIETY SYMPOSIUM PROCEEDINGS

Volume 1536 — Film Silicon Science and Technology, 2013, P. Stradins, A. Boukai, F. Finger, T. Matsui, N. Wyrsch, ISBN 978-1-60511-513-9

Volume 1537E — Organic and Hybrid Photovoltaic Materials and Devices, 2013, S.W. Tsang, ISBN 978-1-60511-514-6

Volume 1538 — Compound Semiconductors: Thin-Film Photovoltaics, LEDs, and Smart Energy Controls, 2013, M. Al-Jassim, C. Heske, T. Li, M. Mastro, C. Nan, S. Niki, W. Shafarman, S. Siebentritt, Q. Wang, ISBN 978-1-60511-515-3

Volume 1539E — From Molecules to Materials—Pathways to Artificial Photosynthesis, 2013, J.-H. Guo, ISBN 978-1-60511-516-0

Volume 1540E — Materials and Integration Challenges for Energy Generation and Storage in Mobile Electronic Devices, 2013, M. Chhowalla, S. Mhaisalkar, A. Nathan, G. Amaratunga, ISBN 978-1-60511-517-7

Volume 1541E — Materials for Vehicular and Grid Energy Storage, 2013, J. Kim, ISBN 978-1-60511-518-4

Volume 1542E — Electrochemical Interfaces for Energy Storage and Conversion—Fundamental Insights from Experiments to Computations, 2013, J. Cabana, ISBN 978-1-60511-519-1

Volume 1543 — Nanoscale Thermoelectric Materials, Thermal and Electrical Transport, and Applications to Solid-State Cooling and Power Generation, 2013, S.P. Beckman, H. Böttner, Y. Chalopin, C. Dames, P.A. Greaney, P. Hopkins, B. Li, T. Mori, T. Nishimatsu, K. Pipe, R. Venkatasubramanian, ISBN 978-1-60511-520-7

Volume 1544E — *In-Situ* Characterization Methods in Energy Materials Research, 2014, J.D. Baniecki, P.C. McIntyre, G. Eres, A.A. Talin, A. Klein, ISBN 978-1-60511-521-4

Volume 1545E — Materials for Sustainable Development, 2013, R. Pellenq, ISBN 978-1-60511-522-1

Volume 1546E — Nanoparticle Manufacturing, Functionalization, Assembly and Integration, 2013, H. Fan, T. Hyeon, Z. Tang, Y. Yin, ISBN 978-1-60511-523-8

Volume 1547 — Solution Synthesis of Inorganic Functional Materials—Films, Nanoparticles and Nanocomposites, 2013, M. Jain, Q.X. Jia, T. Puig, H. Kozuka, ISBN 978-1-60511-524-5

Volume 1548E — Nanomaterials in the Subnanometer-Size Range, 2013, J.S. Martinez, ISBN 978-1-60511-525-2

Volume 1549 — Carbon Functional Nanomaterials, Graphene and Related 2D-Layered Systems, 2013, P.M. Ajayan, J.A. Garrido, K. Haenen, S. Kar, A. Kaul, C.J. Lee, J.A. Robinson, J.T. Robinson, I.D. Sharp, S. Talapatra, R. Tenne, M. Terrones, A.L. Elias, M. Paranjape, N. Kharche, ISBN 978-1-60511-526-9

Volume 1550E — Surfaces of Nanoscale Semiconductors, 2013, M.A. Filler, W.A. Tisdale, E.A. Weiss, R. Rurali, ISBN 978-1-60511-527-6

Volume 1551 — Nanostructured Semiconductors and Nanotechnology, 2013, I. Berbezier, J-N. Aqua, J. Floro, A. Kuznetsov, ISBN 978-1-60511-528-3

Volume 1552 — Nanostructured Metal Oxides for Advanced Applications, 2013, A. Vomiero, F. Rosei, X.W. Sun, J.R. Morante, ISBN 978-1-60511-529-0

Volume 1553E — Electrical Contacts to Nanomaterials and Nanodevices, 2013, F. Léonard, C. Lavoie, Y. Huang, K. Kavanagh, ISBN 978-1-60511-530-6

Volume 1554E — Measurements of Atomic Arrangements and Local Vibrations in Nanostructured Materials, 2013, A. Borisevich, ISBN 978-1-60511-531-3

Volume 1556E — Piezoelectric Nanogenerators and Piezotronics, 2013, X. Wang, C. Falconi, S-W. Kim, H.A. Sodano, ISBN 978-1-60511-533-7

Volume 1557E — Advances in Scanning Probe Microscopy for Imaging Functionality on the Nanoscale, 2013, S. Jesse, H.K. Wickramasinghe, F.J. Giessibl, R. Garcia, ISBN 978-1-60511-534-4

Volume 1558E — Nanotechnology and Sustainability, 2013, L. Vayssieres, S. Mathur, N.T.K. Thanh, Y. Tachibana, ISBN 978-1-60511-535-1

Volume 1559E — Advanced Interconnects for Micro- and Nanoelectronics—Materials, Processes and Reliability, 2013, E. Kondoh, M.R. Baklanov, J.D. Bielefeld, V. Jousseaume, S. Ogawa, ISBN 978-1-60511-536-8

Volume 1560E — Evolutions in Planarization—Equipment, Materials, Techniques and Applications, 2013, C. Borst, D. Canaperi, T. Doi, J. Sorooshian, ISBN 978-1-60511-537-5

Volume 1561E — Gate Stack Technology for End-of-Roadmap Devices in Logic, Power and Memory, 2013, S. Banerjee, ISBN 978-1-60511-538-2

MATERIALS RESEARCH SOCIETY SYMPOSIUM PROCEEDINGS

Prior Materials Research Symposium Proceedings available by contacting Materials Research Society

Thin Film Silicon Solar Cells

Mater. Res. Soc. Symp. Proc. Vol. 1536 © 2013 Materials Research Society
DOI: 10.1557/opl.2013.745

Light Management Using Periodic Textures for Enhancing Photocurrent and Conversion Efficiency in Thin-Film Silicon Solar Cells

Hitoshi Sai[1], Takuya Matsui[1], Adrien Bidiville[1], Takashi Koida[1], Yuji Yoshida[1], Kimihiko Saito[2], Michio Kondo[1]

[1] Research Center for Photovoltaic Technologies, National Institute of Advanced Industrial Science and Technology, Central 2, Umezono 1-1-1, Tsukuba 305-8568, Japan.

[2] Tsukuba Research Laboratory, Photovoltaic Power Generation Technology Research Association, Central 2, Umezono 1-1-1, Tsukuba 305-8568, Japan.

ABSTRACT

Periodically textured back reflectors with hexagonal dimple arrays are applied to thin-film microcrystalline silicon (μc-Si:H) solar cells for enhancing light trapping. The period and aspect ratio of the honeycomb textures have a big impact on the photovoltaic performance. When the textures have a moderate aspect ratio, the optimum period for obtaining a high short circuit current density (J_{SC}) is found to be equal to or slightly larger than the cell thickness. If the cell thickness exceeds the texture period, the cell surface tends to be flattened and texture-induced defects are generated, which constrain the improvement in J_{SC}. Based on these findings, we have fabricated optimized μc-Si:H cells achieving a high active-area efficiency exceeding 11% and a J_{SC} of 30 mA/cm^2.

INTRODUCTION

Thin-film silicon solar cells (TFSSC) are a promising candidate for the future large-area photovoltaic systems operating in the gigawatt scale in high-temperature regions, because of the abundance and non-toxicity of the source materials and the superior temperature coefficient [1-2]. However, further conversion efficiency improvement is desirable to make TFSSC more cost-competitive. In TFSSC, so-called light trapping technology is crucial to absorb photons within thin Si films to compensate for their insufficient carrier transport properties and the high film deposition cost especially for microcrystalline silicon (μc-SI:H). For this purpose, textured substrates have been implemented to scatter the incident light and increase the optical path length inside the cells [1-22].

In recent years, periodically textured substrates or surface gratings have been actively studied as a more sophisticated platform with a potential for realizing a higher current density than achievable with the conventional random textures [9-22]. Because of their simplicity and uniformity in texture morphology, periodic structures have the possible advantage of much clearer correlations between texture parameters and the photovoltaic performance in solar cells. In addition, the periodicity allows us to use periodic boundary conditions in optical simulations, which reduces the calculation cost substantially. The high potential of periodic textures has been demonstrated in hydrogenated amorphous Si (a-Si:H) solar cells [9]. More recently, our group reported that a carefully chosen periodic texture with hexagonal dimple array (honeycomb texture) enhances the conversion efficiency as well as the current density in μc-Si:H cells [14-16]. In this report, we review our recent works on light trapping in μc-Si:H cells with honeycomb textures. We show that the photovoltaic performance is strongly influenced by structural parameters of the textures. Based on our findings, we optimize the periodic textures for each cell thickness. Accordingly, a high short circuit current density (J_{SC}) exceeding 30 mA/cm^2 and a high efficiency of 11.1% are obtained in μc-Si:H cells with an active area of 1 cm^2.

EXPERIMENT

Honeycomb textures were fabricated by the following procedure. First, a hexagonal photo-resist pattern was formed on a Si wafer with a thermally grown SiO$_2$ film using photolithography. Next, the wafer was soaked in a diluted buffered HF (BHF) solution to transfer the hexagonal pattern onto the SiO$_2$ film. After the removal of the remaining resist, a Ag/ZnO:Ga stacked film (200 nm/100 nm) was deposited on the honeycomb-textured SiO$_2$ layer by sputtering, to obtain a highly reflective and conductive surface. In this work, we used honeycomb substrates with periods of $P = 1.0$ to 4.0 μm and aspect ratios of $H/P = 0$ to 0.3, where H denotes the peak height of the texture.

Substrate-type n-i-p μc-Si:H cells with an active area of 1 cm^2 were fabricated on these substrates. The structure of the solar cell consists of honeycomb-textured substrate/Ag/ZnO/μc-Si:H n-i-p layers/In$_2$O$_3$:Sn(ITO, 70 nm)/Ag grid. The intrinsic μc-Si:H layer was deposited with conventional plasma-enhanced chemical vapor deposition (PECVD) using SiH$_4$/H$_2$ gas mixture, with the thickness t_i varying from 0.5 to 3 μm. The n- and p-type μc-Si:H were also deposited by PECVD using PH$_3$ and B$_2$H$_6$ as dopant gases, respectively. For some solar cells, wide-gap nanocrystalline silicon oxides (nc-SiO$_X$:H) were applied in the doped layers for reducing the absorption loss [5]. The top ITO film and Ag finger grid electrodes were deposited by sputtering at room temperature. Isolation of the cells was done by reactive ion etching. Finally, all the cells were annealed at 175°C for 2 h. The performance of the cells was evaluated by measuring current-voltage (J-V) characteristics with a dual-light solar simulator under AM 1.5G and 100 mW/cm^2 and external quantum efficiency (EQE) spectra. Reflectivity of the cells was also measured using a spectrometer with an integrating sphere (Perkin Elmer, Lambda 950).

RESULTS AND DISCUSSION

Morphology of substrates and μc-Si:H cells

Figure 1 shows scanning electron microscope (SEM) images of μc-Si:H solar cell with t_i = 1 μm deposited on a honeycomb-textured substrate. Fig. 1 (a) illustrates a uniform honeycomb texture with dimples fabricated by the procedure mentioned above. As seen in Fig. 1 (b), the morphology of the cell surface is not a simple replication of the substrate. The diameter of each dimple is reduced after μc-Si:H film growth, and shrunk holes are formed on the cell surface. This can be also confirmed in Figs. 1 (c, d), showing cross-sectional views of the cell. Fig. 1 (c) shows a cross section along the line A, that includes the ridges of honeycombs. The sides of the

4

holes in each pattern can be tapered in a manner similar to a crater, as shown in Fig. 1 (c), by properly choosing the composition of BHF solution and the dipping time. This feature is advantageous to avoid the generation of defects, which are often grown at steep valleys on textured substrates [23, 24]. It is also seen that the height of the texture varies along the ridge, and the peak height, H, is obtained at the crossing point of the ridge lines. A similar height variation is also found on the cell surface. On the other hand, Fig. 1 (d) shows a cross section along the line B, that passes through the centers of adjacent dimples. In this figure, the spacing of the texture peaks corresponds to the period, P. The diameter of the flat area at the bottom of each dimple, D, is also controllable. In this study, D/P is ranged from 0.3 to 0.5.

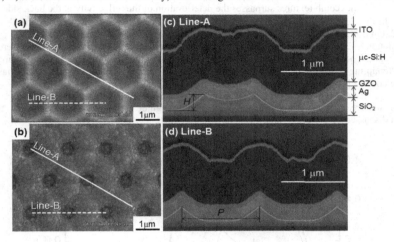

Figure 1. SEM images of a 1-μm-thick μc-Si:H solar cell fabricated on a honeycomb-textured substrate with $P = 1.5$ μm and $H/P = 0.23$: top views of (a) the substrate and (b) the solar cell, cross-sections of the cell along (c) the line A and (d) the line B.

Effects of texture parameters in 1-μm-thick cells

Figure 2 shows the measured EQE spectra of the μc-Si:H solar cells fabricated on the honeycomb-textured substrates with different H/P at an almost constant period of $P = 1.4 - 1.5$ μm. The cell thickness was fixed at $t_i = 1$ μm in this figure. The EQE is enhanced in the long wavelength region by the introduction of the honeycomb textures with higher aspect ratios. A high current density of 25.1 mA/cm^2 is attained at $(P, H/P) = (1.5$ μm, 0.23), which is higher by 7.6 mA/cm^2 (+43%) with respect to the flat cell. In addition, the EQE is also improved in the wavelengths shorter than 500 nm, which is ascribed to the anti-reflection effect by the textured front surface of the cells. However, the EQE peaks of the textured cells are slightly lower than that of the flat cell, suggesting that the reflectivity at the back side is decreased by the honeycomb textures. To confirm this, we have calculated the reflectivity of the honeycomb texture with $(P, H/P) = (1.5 \ \mu m, 0.23)$ in a μc-Si:H cell by using finite-difference time-domain (FDTD) method [25]. A three-dimensional calculation model has been developed based on the cross-sectional SEM images as shown in Fig. 1. In the calculation, optical constants (n, k) measured in our laboratory or reported in the literature [26, 27] were used, although extinction coefficients of μc-Si:H are assumed to be 0 $(k = 0)$ for all wavelengths due to the restriction of our calculation program. As can be seen in Fig. 3, the honeycomb texture shows lower reflectivity than a flat Ag/ZnO stack across the whole wavelength range. A notable absorption peak observed at 550 nm is ascribed to surface plasmon on the textured Ag surface. Nevertheless, the substantial enhancement in EQE obtained here indicates that the improvement in light trapping by honeycomb textures surpasses the deterioration of the reflectivity at the back side.

In Fig. 2, the absorption spectrum of the cell with $(P, H/P) = (1.5 \ \mu m, 0.23)$ is also plotted. On this spectrum, a number of fine peaks, which are not clearly seen in the corresponding EQE spectrum, can be observed in addition to a series of fringes by Fabric-Perot interference. These absorption peaks seem to be signatures of the light trapping in this system: optical coupling of the incident light into guided modes in the cell at specific wavelengths via a grating coupler [10]. However, it's rather difficult to assign each peak on this spectrum to a specific guided mode, because both of diffraction modes generated by the two-dimensional periodic texture and guided modes inside the Si slab waveguide are numerous in the optical system discussed here. Further analysis is now undergoing.

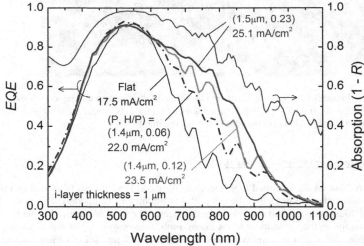

Figure 2. External quantum efficiency (EQE) spectra of 1-μm-thick μc-Si:H cells fabricated on a flat substrate and honeycomb-textured substrates with (P, H/P) = (1.4 μm, 0.06), (1.4 μm, 0.12), and (1.5 μm, 0.23). The absorption spectrum of the cell with (P, H/P) = (1.5 μm, 0.23) is also plotted.

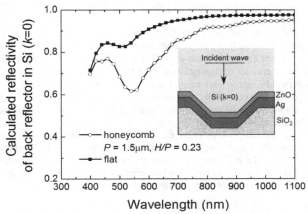

Figure 3. Calculated reflectivity spectra of a honeycomb textured back reflector with $(P, H/P) =$ (1.5 μm, 0.23) and a flat reflector. The inset shows a schematic model used in the calculation. The back reflector contact with pseudo silicon with $k = 0$.

In Figure 4, J-V parameters of the 1-μm-thick μc-Si:H solar cells on the honeycomb textures with $P = 1.4$ and 1.5 μm are plotted as functions of H/P. As can be seen in Fig. 4 (a), the short-circuit current density J_{SC} increases drastically with increasing H/P and then saturates at $H/P \approx 0.23$. The maximum values of $J_{SC} = 25.5$ mA/cm^2 was obtained for the cells with the standard (p)μc-Si:H. However, J_{SC} begins to decrease at higher H/P. It was confirmed by EQE measurement under reverse bias voltages that the cells at higher H/P have negligible carrier collection loss. On the other hand, the reflectivity of the honeycomb back reflectors measured in air decreases gradually with increasing H/P, which also leads the reduced reflectivity in the device, as seen in Fig. 3. Therefore, we ascribe the decrease in J_{SC} at higher H/P to the enhanced absorption loss mainly caused by the reduced reflectivity of the back reflectors. The open-circuit voltage (V_{OC}) shows a slight decrease as H/P increases, whereas the fill factor (FF) is rather stable, as plotted in Figs. 4 (b) and (c). The decreasing trend in V_{OC} can be ascribed to the insufficient coverage by the doped layers and/or the deterioration in the Si film quality due to the rougher growth surface. Overall, the cell performance is mainly governed by J_{SC}, and a markedly high efficiency of 10.1% is achieved at $H/P \approx 0.23$, as shown in Fig. 4 (d).

Light trapping in μc-Si:H cells is also influenced substantially by the period of the honeycomb textures. In Figure 5, J-V parameters are plotted as functions of P under a fixed aspect ratio of $H/P \sim 0.2$. It is clearly shown in Fig. 5 (a) that the optimum period for getting a high J_{SC} is around $P = 1.5$ μm in the case of $t_i = 1$ μm, while a high J_{SC} of over 24 mA/cm^2 is achievable when $P = 1 - 1.5$ μm. Conversion efficiency is mainly governed by J_{SC} and the highest efficiency is attained at $P = 1.5$ μm. Relatively minor variations in V_{OC} and FF observed in Figs. 5 (b) and (c) reveal that honeycomb-textured substrates used here do not deteriorate the quality of μc-Si:H layers, at least in the case of $t_i = 1$ μm. Similarly to the case of H/P, therefore, the conversion efficiency is mainly dominated by the J_{SC} and the maximum value is obtained at $P = 1.5$ µm, as shown in Fig. 5 (d).

To date, several groups have analyzed light trapping in thin-film silicon solar cells by using optical calculation tools such as FDTD, finite integration technique (FIT), and finite element method (FEM). For example, Haase et al. [17] and Dewan et al. [18] reported that, in µc-Si:H cells with the superstrate configuration and $t_i = 1$ µm, optimal periods and heights of three-dimensional pyramidal textures are ranged in $P = 0.7 - 1.2$ µm and $H = 0.3 - 0.5$ µm, respectively. When comparing our results with theirs, it is found that the optimal P is slightly longer than what is simulated, while the optimal H, that is 1.5 µm $\times$ 0.23 = 0.35 µm, is within the simulated range. The discrepancy between the simulations and our results seems to be originated from the deformation of the morphology of the cell surface during Si film growth. In the simulations, they assume the exactly same surface morphology on both of the front and back surfaces. However, in real solar cells, surface texture on the Si film changes gradually during film growth, as seen in Fig. 1, and the growth surface tends to be flatter than the original texture in most cases. We will discuss this issue in the following section in more detail.

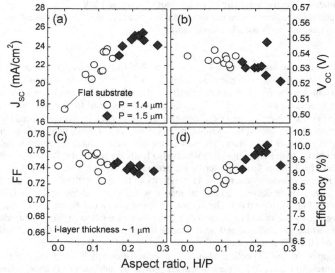

Figure 4. J-V parameters of µc-Si:H solar cells fabricated on honeycomb textures with periods of $P = 1.4$ µm or 1.5 µm as functions of the aspect ratio, H/P. The cell thickness is fixed at $t_i = 1$ µm.

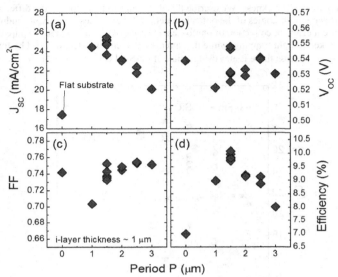

Figure 5. *J-V* parameters of 1-μm-thick μc-Si:H solar cells fabricated on honeycomb textures with aspect rations of $H/P \sim 0.2$ as functions of the period, *P*.

Interplay between texture period and cell thickness

In the previous section, μc-Si:H cells with a fixed thickness of $t_i = 1$ μm have been investigated. However, the texture period giving the highest J_{SC} is dependent on the cell thickness. Figure 6 shows the relationship between the period of honeycomb-textured substrates and the J_{SC} of μc-Si:H solar cells with $t_i = 0.5, 1, 2,$ and 3 μm. Here we fixed H/P in the range 0.2–0.25. The J_{SC} obtained by using quasi-periodically textured Al substrates [12] are plotted as supplemental data in the range of $P < 1$ μm. As shown in Fig. 6, the optimum period providing the highest J_{SC} becomes longer with increasing t_i. In the case of $t_i = 0.5$ μm, the optimum period is found to be around 1 μm, although a clear peak is not observable due to the lack of experimental data in the range $P < 1$ μm. A high J_{SC} of over 30 mA/cm^2 is attained at $P = 3$ to 4 μm for cells with $t_i = 3$ μm and a (p)nc-SiO$_X$ layer. This value is comparable to the highest-current cell with state-of-the-art random textures [7].

Recent theoretical considerations have shown that optimal light trapping occurs when the grating period approximately equals to the absorption edge wavelength of the material, which is ~1 μm in the case of μc-Si:H [21, 22]. Note that this conclusion is deduced assuming that incident light behaves incoherently, while we observed interference fringes in EQE and absorption spectra, as shown in Fig. 2. As mentioned previously, this has been supported by several optical modeling results [17–19]. Based on this theory, the optimum period is independent of the thickness of solar cells. However, we have found a clear dependence of the optimal period on the cell thickness. We ascribe this difference to two aspects: the difference between the surface morphologies of the optical models and the devices, and texture-induced defects that prevent efficient collection of photo-generated carriers. The former is a pure optical effect directly linked to light trapping, while the latter is an electrical effect induced by textures. In the following, we discuss these issues using a pair of μc-Si:H cells as an example.

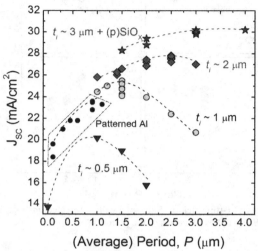

Figure 6. The relationship between the period of textured substrates and the J_{SC} of μc-Si:H solar cells with i-layer thicknesses of $t_i = 0.5$, 1, 2, and 3 μm. For cells with $t_i = 3$ μm, a wide-gap (p)nc-SiO$_X$ layer was applied. Circles surrounded by dashed lines represent data obtained with 1-μm-thick cells on patterned Al substrates fabricated by anodic oxidation (see Ref. 12). Dashed lines in this figure are guides to the eye.

In Fig. 7, we compare the EQE and absorption spectra of the 3-μm-thick μc-Si:H cells fabricated on honeycomb textures with $P = 1.5$ and 3 μm. First we focus on the absorption spectra to discuss the optical effect. As shown in Fig. 7, the honeycomb texture with $P = 3.0$ μm exhibits higher absorption than the other across almost the entire wavelength region shown here, in contradiction to the theoretical prediction. In optical modeling, an ideal texture is often assumed on both sides of the solar cells. However, in conventional TFSSCs, the morphology of the growth surface is not controlled, resulting in a complex morphology due to flattening and self-texturing during film growth. In fact, as seen in Figs. 1 (b), the dimples on the cell surface become smaller than those on the substrate, which results in a flatter cell surface and insufficient in-coupling and trapping of the incident light. An EQE enhancement by front texturing has been experimentally confirmed using μc-Si:H cells with intentionally polished surfaces [8]. In principle, the grating period should be larger than the cell thickness to avoid severe flattening of the cell surface. Therefore, in TFSSCs with an as-deposited surface morphology, the optimum period should be larger than the cell thickness.

10

Next we move to the electrical effect. As shown in Fig. 7, the honeycomb texture with P = 3.0 μm exhibit a higher EQE than that with P = 1.5 μm. The difference in the EQE is more pronounced than that in absorption spectra. A reverse bias voltage applied during the EQE measurement especially helps the carrier collection in the case of P = 1.5 μm, as plotted in Fig. 7. This result indicates that, for thick cells, the texture with P = 1.5 μm induces more carrier recombination centers than the other. Figure 8 shows cross sectional bright-field scanning electron microscope (STEM) images of the cells. As shown in Fig. 8 (a), in the case of P = 1.5 μm, bright streaky parts can be observed underneath the concave regions on the cell surface. The bright streak, which is composed of a low-density amorphous structure, starts to grow roughly in the middle of the cell and ends with a void at the cell surface. By comparing Fig. 8 (a) and Fig. 1 (d), the evolution of the growth surface can be estimated as schematically drawn in Fig. 8 (a) with white dashes lines. The flat area at the bottom of the concave is expected to become smaller with increasing thickness, and to disappear completely at a certain thickness, which probably leads insufficient supply of Si precursors (shadowing effect) and makes a low-density part. The low-density part is essentially identical to the so-called "crack" reported by other groups [23, 24]. Moreover, the non-uniform coverage of the p-layer around the concaves could weaken the internal electric field. All these effects can cause insufficient carrier collection and a reduction in J_{SC}. In contrast, in the case of P = 3 μm, no clear defective region is found within the entire μc-Si:H layer, as is evident from Fig. 8 (b). This result confirms that the enlarged dimples can mitigate the generation of texture-induced defects and resulting carrier collection problems in thick cells. Note that the surface flattening of the cell on the texture with P = 1.5 μm is also observable when comparing Figs. 1 (d) and Fig. 8 (a).

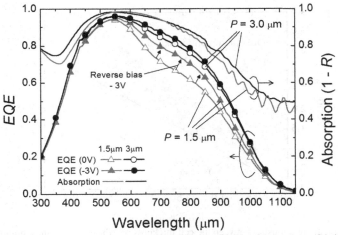

Figure 7. EQE and absorption (1-R) spectra of μc-Si:H solar cells with t_i = 3 μm fabricated on honeycomb-textured substrates with P = 1.5 μm (grey lines with triangles) and 3.0 μm (black lines with circles and solid line).

To summarize the finding mentioned above, the relationship between the cell thickness and the period giving the highest J_{SC} at each thickness is schematically shown in Fig. 9. We have found that the optimal period for highest J_{SC} in TFSSC is not constant but depends strongly on the cell thickness; the optimal period should be equal or slightly larger than the cell thickness. In addition, the effect of the period on J_{SC} becomes weaker with increasing the cell thickness. These results show that the cell thickness must be taken into account to design optimum periodic surface textures for light trapping in TFSSCs, at least when using moderate aspect ratios.

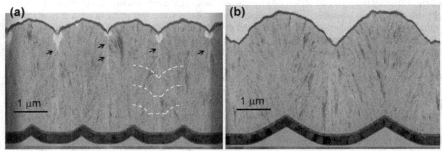

Figure 8. Bright-field STEM images of μc-Si:H cells on honeycomb-textured substrates with (a) $P = 1.5$ μm and (b) $P = 3$ μm. The thickness of the cells is $t_i = 3$ μm. Black arrows in (a) indicate low-density defective parts that exist in the upper half of the cell. White dashed lines show schematic cell surfaces during film growth.

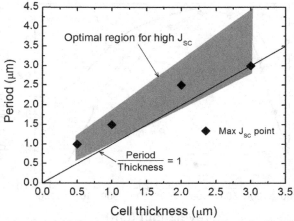

Figure 9. Schematic relationship between the cell thickness and the optimal honeycomb period for achieving a high J_{SC}. Solid circles correspond to the highest J_{SC} points at some thicknesses shown in Fig. 6. The grey area indicates the region where a high $J_{SC} > (0.98 \times$ maximum $J_{SC})$ has been obtained in this study.

Finally, the J-V properties of the most efficient µc-Si:H solar cells fabricated in this study are summarized in Table I. By using properly chosen honeycomb textures with respect to the cell thickness and more transparent doped layers, we have realized a 2-µm-thick cell with an active area efficiency of 11.1%. This is among the highest efficiencies reported thus far for single-junction µc-Si:H cells, demonstrating a high potential advantage for periodic textures.

TABLE I. J-V parameters of µc-Si:H cells fabricated on a flat and honeycomb texture. All the cells have an active area of 1 cm^2.

Substrate	i-layer thickness (µm)	p-layer	n-layer	V_{OC} (V)	J_{SC} (mA/cm^2)	FF	Eff.
Flat	1	µc-Si:H	µc-Si:H	0.539	17.5	0.742	7.0%
P = 1.5 µm	1	SiO$_X$	µc-Si:H	0.529	26.3	0.739	10.3%
P = 2.5 µm	2	SiO$_X$	SiO$_X$	0.523	28.6	0.741	11.1%

CONCLUSIONS

We have shown photocurrent and efficiency improvements in µc-Si:H cells deposited on back reflectors with periodic honeycomb textures. The effects of the texture parameters such as the period and the aspect ratio on the photovoltaic performance have been investigated in detail. It is found that the optimum period for obtaining a high J_{SC} as well as high efficiency requires being equal to or slightly larger than the cell thickness. This can be explained by the following mechanisms: When the cell thickness exceeds the texture period, J_{SC} is limited by (i) deteriorated light trapping in the cell due to the surface flattening, and (ii) generation of texture-induced defects. The latter is probably caused by shadowing effect during Si deposition. If the texture period is excessively long, the light-trapping effect becomes weaker and the J_{SC} is decreased, as predicted by optical simulations. We emphasize here that such clear correlations between the cell thickness and the texture period are obtained by utilizing periodic structures. Based on the findings described above, we have optimized the honeycomb texture and achieved an active area efficiency of 11.1% with t_i = 2 µm and a J_{SC} of 30 mA/cm^2 with t_i = 3 µm, indicating a high potential advantage for periodic textures. Application of theoretically optimum textures to TFSSC would require the use of new approaches, such as those involving flattened substrates [28, 29] or post-deposition texturing, to avoid constraints that normally accompany the growth of Si films on textured substrates. Furthermore, the application of our results to the superstrate configuration, which is a dominant architecture in current TFSSCs, also remains as a future task.

ACKNOWLEDGMENTS

The authors thank Ms. N. Hozuki for her help in this work. A part of this work was supported by New Energy and Industrial Technology Development Organization (NEDO), Japan. Photolithography process was conducted at the AIST Nano-Processing Facility, supported by "Nanotechnology Network Japan" of the Ministry of Education, Culture, Sports, Science and

Technology (MEXT), Japan.

REFERENCES

1. J. Meier, S. Dubail, R. Platz, P. Torres, U. Kroll, J.A. Anna Selvan, N. Pellaton Vaucher, Ch. Hof, D. Fischer, H. Keppner, R. Flückiger, A. Shah,V. Shklover, K.-D. Ufert, Sol. Energy Mater. Sol. Cells **49**, 35 (1997).
2. K. Yamamoto, T. Suzuki, M. Yoshimi, A. Nakajima, Jpn. J. Appl. Phys. **36**, L569 (1997).
3. M. Kambe, A. Takahashi, N. Taneda, K. Masumo, T. Oyama, and K. Sato, *Proc. 33rd IEEE Photovolt. Specialist Conf.*, San Diego, 2008, pp.609-613.
4. W.W. Wenas, A. Yamada, M. Konagai, K. Takahashi, Jpn. J. Appl. Phys. **30** L441 (1991).
5. M. Despeisse, C. Battaglia, M. Boccard, G. Bugnon, M. Charrière, P. Cuony, S. Hänni, L. Löfgren, F. Meillaud, G. Parascandolo, T. Söderström, C. Ballif, Phys. Status Solidi A **208**, 1863 (2011).
6. M. Berginski, J. Hüpkes, M. Schulte, G. Schöpe, H. Steibig, B. Rech, and M. Wuttig, J. Appl. Phys. **101** 074903 (2007).
7. B. Yan, G. Yue, L. Sivec, J. Owens-Mawson, J. Yang, S. Guha, Sol. Energy Mater. Sol. Cells **104**, 13 (2012).
8. H. Sai, H. Jia, M. Kondo, J. Appl. Phys. **108**, 044505 (2010).
9. C. Battaglia, C.-M. Hsu, K. Söderström, J. Escarre, F.-J. Haug, M. Charrière, M. Boccard, M. Despeisse, D. T. L. Alexander, M. Cantoni, Y. Cui, C. Ballif, ACS Nano **6**, 2790 (2012).
10. F.-J. Haug, K. Söderstrom, A. Naqavi, C. Ballif, J. Appl. Phys. **109**, 084516 (2011).
11. F.-J. Haug, T. Söderström, M. Python, V. Terrazzoni-Daudrix, X. Niquille, C. Ballif, Sol. Energy Mater. Sol. Cells **93**, 884 (2009).
12. H. Sai, M. Kondo, J. Appl. Phys. **105**, 094511 (2009).
13. M. Vanecek, O. Babchenko, A. Purkrt, J. Holovsky, N. Neykova, A. Poruba, Z. Remes, J. Meier, U. Kroll, Appl. Phys. Lett. **98**, 163503 (2011).
14. H. Sai, K. Saito, M. Kondo, Appl. Phys. Lett **101**, 173901 (2012).
15. H. Sai, K. Saito, M. Kondo, IEEE J. Photovolt. **3**, 5 (2013).
16. H. Sai, K. Saito, N. Hozuki, M. Kondo, Appl. Phys. Lett **102**, 053509 (2013).
17. C. Haase and H. Stiebig, Appl. Phys. Lett. **91**, 061116 (2007).
18. R. Dewan, I. Vasilev, V. Jovanov, D. Knipp, J. Appl. Phys. **110**, 013101 (2011).
19. A. Čampa, Janez Krč, M. Topič, J. Appl. Phys. **105**, 083107 (2009).
20. X. Sheng, J. Liu, I. Kozinsky, A.M. Agarwall, J. Michel, L.C. Kimerling, Adv. Mater. **23**, 843 (2011).
21. Z. Yu, A. Raman, S. Fan, Opt. Express **18**, A366 (2010).
22. S. E. Han, G. Chen, Nano Lett. **10**, 4692 (2010).
23. M. Python, O. Madani, D. Dominé, F. Meillaud, E. Vallat-Sauvain, C. Ballif, Sol. Energy Mater. Sol. Cells **93**, 1714 (2009).
24. H.B.T. Li, R.H. Franken, J.K. Rath, R.E.I. Schropp, Sol. Energy Mater. Sol. Cells **93**, 338-349 (2009).
25. For example, A. Taflove, S.C. Hagness, Computational Electrodynamics – finite-difference time-domain method, (Artech House, 2000).
26. D.W. Lynch, W.R. Hunter, Handbook of Optical Constants of Solids, ed. E.D. Palik (Academic Press, 1985) pp.351-357.

27. H. Fujiwara, M. Kondo, Phys. Rev. B **71**, 075109 (2005).
28. H. Sai, Y. Kanamori, M. Kondo, Appl. Phys. Lett. **98**, 113502 (2011).
29. K. Söderstrom, G. Bugnon, R. Biron, C. Pahud, F. Meillaud, F.-J. Haug, C. Ballif, J. Appl. Phys. **112**, 114503 (2012).

Mater. Res. Soc. Symp. Proc. Vol. 1536 © 2013 Materials Research Society
DOI: 10.1557/opl.2013.596

Panasonic's Thin Film Silicon Technologies for Advanced Photovoltaics

Akira Terakawa, Hiroko Murayama, Yohko Naruse, Hirotaka Katayama, Takeyuki Sekimoto, Shigeo Yata, Mitsuhiro Matsumoto, Isao Yoshida, Mitsuoki Hishida, Youichiro Aya, Masahiro Iseki, Mikio Taguchi and Makoto Tanaka

Panasonic Corporation
3-1-1 Yagumo-naka-machi, Moriguchi 570-8501, Japan

ABSTRACT

We have fabricated high-efficiency a-Si/μc-Si tandem solar cells and modules with a very high μc-Si deposition rate using Localized Plasma Confinement CVD to give very high-rate deposition (>2.0 nm/s) of device-grade μc-Si layers. For further progress in productive plasma-CVD techniques, we have studied plasma phenomena by combining newly developed plasma simulation and plasma diagnosis techniques that reveal the importance of non-emissive atomic hydrogen. We also have proposed a model of defective μc-Si formation on highly textured substrates in which the atomic H in plasma is assumed to play an important role. We are also developing a non-vacuum deposition technique that we term "Liquid Si Printing." A new record conversion efficiency for HIT solar cells of 24.7% has been achieved using a very thin c-Si wafer (Thickness: 98 μm, Area: 102 cm^2).

INTRODUCTION

Thin-film silicon, such as hydrogenated amorphous silicon (a-Si), microcrystalline silicon (μc-Si) and related alloys, are promising materials for very low-cost solar cells. Sanyo has conducted R&D on thin-film silicon solar cells for more than 30 years, and released the first a-Si solar cell products, the AMORTON series, in 1980, and the highest-performing photovoltaic modules, called the HIT Power Series, in 1997. Sanyo's R&D history has been grafted onto Panasonic's. This paper reviews recent progress in thin film silicon technologies for advanced solar cells at Panasonic.

EFFICIENCY RECORDS OF THIN FILM SILICON SOLAR CELLS

We have continuously proposed innovative technologies and achieved record efficiency values for various types of Si-based solar cells (Table I) [1-22]. In 1993, the world's highest initial conversion efficiency of over 12% was achieved for an a-Si single junction solar cell (100 x 100 mm) [1]. In 2001, the world's highest stabilized conversion efficiency of 10% was also achieved for an a-Si/a-SiGe tandem module (900 x 900 mm) with a deposition rate of over 0.2 nm/s [6, 7] for the bottom a-SiGe layer. Replacing the bottom a-SiGe layer with μc-Si effectively improves long-wavelength sensitivity and conversion efficiency. However, a relatively thick μc-Si layer (around 2.0 μm) could be a bottleneck in the manufacturing process. Much effort has been devoted to increasing both the stabilized efficiency and the process throughput of μc-Si deposition. In 2011, we were able to improve the performance of large-size tandem modules. Figure 1 shows an a-Si/μc-Si tandem solar module (G5 size). An initial module efficiency of 12.0% and a stabilized module efficiency of 10.7% were achieved [16-22]. The initial conversion efficiency of 13.5% (Voc: 1.41V, Jsc: 12.7mA/cm^2, FF: 0.754) and a stabilized conversion efficiency of 12.2% have been achieved for a small cell (Area: 1 cm^2) [16-22]. These stabilized efficiencies were the world's highest level for a-Si/μc-Si tandem solar cells.

Table I. Efficiency of Panasonic's thin-film silicon solar cells

Year	Size (cm^2)	Structure	Init. Eff.	Stab. Eff.	Red.
1993	1	a-Si	12.7%	—	[1,2]
1993	100	a-Si	12.0%	—	[1,2]
1994	1	a-Si	—	8.9%	[3]
1994	1	a-Si/a-SiGe	11.6%	10.6%	[4]
1997	1200	a-Si/a-SiGe	11.1%	9.5%	[5]
2001	8,252	a-Si/a-SiGe	11.2%	10.0%	[6, 7]
2007	1	a-Si/μc-Si	13.5%	—	[10]
2010	15,400	a-Si/μc-Si	11.1%	10.0%	[13-15]
2011	14,300	a-Si/μc-Si	12.0%	10.7%	[16-22]
2011	1	a-Si/μc-Si	13.5%	12.2%	[16-22]

Figure 1. Large-area a-Si/μc-Si tandem module.

18

ADVANCED TECHNOLOGIES FOR SILICON FILM DEPOSITION

Localized Plasma Confinement CVD (LPC-CVD)

Plasma CVD equipment is the major cost factor in thin-film solar cell manufacturing. To realize cost-effective (high-performance and low-cost) solar cells, very high-throughput technologies are necessary. High-rate deposition of μc-Si layers is a particularly important challenge because they are several times thicker than a-Si layers. Although it has been reported that high-pressure plasma conditions (>200 Pa) are effective for increasing the deposition rate using small-size equipment [23-25], some tradeoffs exist when adapting these conditions to a large-area deposition process [17]. Since 2005, we have been developing a Localized Plasma Confinement (LPC) CVD method (Fig. 2), and have succeeded in achieving large, stable and uniform plasma under very high pressure conditions (>1000 Pa) [8-22].

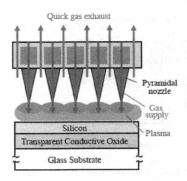

Figure 2. Schematic image of the Localized Plasma Confinement CVD (LPC-CVD) method.

Figure 3 shows trends in the conversion efficiency of μc-Si single and a-Si/μc-Si tandem solar cells deposited by LPC-CVD. The initial efficiency values have improved constantly with increases in the deposition rate of the μc-Si layers and the scaling up of the substrate size. An initial module efficiency of 11.1% and a stabilized module efficiency of 10.0% were finally achieved for large-size (1,100 x 1,400 mm) modules with a deposition rate of 2.4 nm/s, which is one order higher than that of a-SiGe [13-22]. Figure 4 shows the performance of tandem solar cells deposited by the G5.5 size plasma CVD plotted against the deposition rate of μc-Si layers before and after the optimization of process conditions. At first, the conversion efficiency deteriorated with increased deposition rate, as generally reported. However, we were able to achieve high-quality μc-Si and high-efficiency tandem solar cells independent of the deposition rate in the range of 1.0 - 2.5 nm/s after optimizing the plasma CVD conditions under very high-pressure conditions.

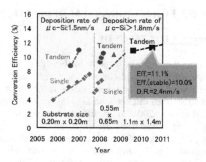

Figure 3 Trends in the conversion efficiency of solar cells, substrate size and deposition rate of the intrinsic layer of µc-Si.

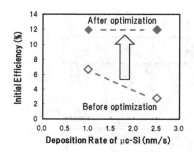

Figure 4 Performance of tandem solar cells deposited by the G5.5 size plasma-CVD. (◇: before optimization, ♦: after optimization).

Radical Flux Analysis

We investigated the effect of SiH$_3$ and atomic hydrogen fluxes (Γ_{SiH3} and Γ_H) on deposition rate by comparing the substrate temperature dependence of deposition rate and those fluxes [14, 21]. We found that (a) Γ_H does not always correspond to Hα emission intensity and (b) non-emissive H is also important, especially under high-pressure deposition conditions. To be able to predict Γ_H for any CVD equipment under various deposition conditions, we developed an original radical flux simulator, named the "high-pressure SiH$_4$ plasma simulator (HiP-SPS)" [21] that is potentially useful for the design of much more productive plasma-CVD equipment and processing conditions.

Defective Regions in Microcrystalline Silicon

For further improvement of conversion efficiency, advanced techniques, such as material design and optical confinement using very high-Haze transparent conductive oxide (TCO), are effective. However, when depositing μc-Si films at very high deposition rates on very high-Haze substrates, including steep texture valleys, defective regions are generated that degrade solar cell performance [26-29]. We tried to reveal the detailed structure of defective regions and have proposed a formation model [20].

Figures 5(a) and 5(b) show SEM and TEM cross-section images of the μc-Si film, respectively. The circle-identified regions (dark regions in Fig. 5(a) and bright regions in Fig. 5(b)) are low-density regions. By slicing the same sample, we were able to identify the microstructure of the defective regions, which consist of a vacancy and low-density a-Si region (around the vacancy) as shown in Figure 6. These microstructures were probably formed due to the lack of H radical flux in the narrow valley of the high-Haze texture, because H radical density decreases more drastically as a result of surface collisions than SiH_3 [20]. The control of H radical flux thus appears to be a key parameter for high-quality μc-Si films with very high-rate deposition on very high-Haze texture.

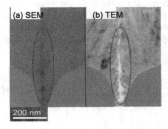

Figure 5 Cross-section images of defective regions in μc-Si at a valley with high-Haze texture: (a) SEM and (b) TEM

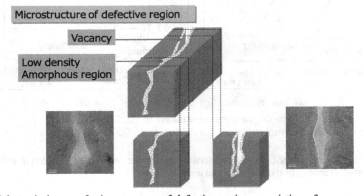

Figure 6 Schematic image of microstructure of defective regions consisting of a vacancy and low-density a-Si.

"LIQUID-SILICON PRINTING," AN ALTERNATIVE TECHNIQUE

Positioned as a future technology, we are developing a non-vacuum "Liquid-Si printing (LSP) method," which consists of printing Liquid-Si and pyrolyzing for solidification. The Liquid-Si contains polydihydrosilane ($-(SiH_2)_n-$), suggested by Professor Tatsuya Shimoda at JAIST (Japan Advanced Institute of Science and Technology) [30, 31]. The printed a-Si:H by LSP were previously demonstrated in TFT fabrication [30]. It was also reported that the printed a-Si:H films shows relatively high photoconductivities ($\sim 1.0 \times 10^{-5}$ S/cm) [31]. This suggests that LSP a-Si could work as a photovoltaic layer in solar cells. For printed a-Si:H with device-grade properties, catalytic-generated atomic hydrogen (Cat-H*) treatment proved effective and necessary. In this study, we analyzed the effect of Cat-H* treatment of a-Si:H printed using LSP.

We showed the Si films and the a-Si solar cells to have extremely good photo-stability [32]. Figures 7(a) and 7(b) show the photo degradation and thermal recovery behaviors characteristic of photo-stability (7(a)) and normalized conversion efficiency (7(b)). The light-induced degradation of the LSP printed a-Si:H is much less than that of the CVD-deposited a-Si:H, although the velocities of thermal recovery of the photoconductivity and the conversion efficiency were comparable. The photo-stability of the LSP printed a-Si:H is probably due to the unique hydrogen bonding structures, which are very different from those of CVD-deposited a-Si:H. These investigations are steps toward achieving fully printed solar cells without the use of any vacuum equipment.

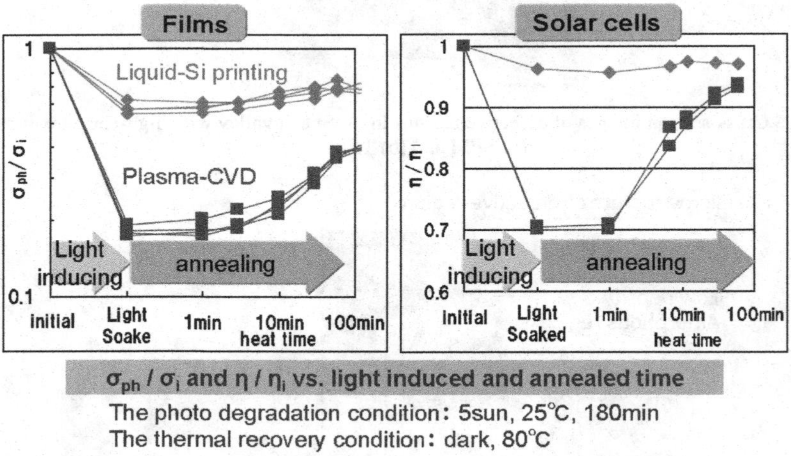

σ_{ph} / σ_i and η / η_i vs. light induced and annealed time

The photo degradation condition: 5sun, 25°C, 180min
The thermal recovery condition: dark, 80°C

Figure 7 (a) the normalized photoconductivities and (b) the normalized conversion efficiencies plotted against light-induced time and annealed time.

HIT SOLAR CELLS

On the basis of a-Si technology using plasma-CVD, a-Si/c-Si heterojunction structure termed HIT (heterojunction with intrinsic thin layer) solar cells were developed in 1990 [33, 34]. HIT solar cells have the following features: (1) excellent surface passivation which results in high voltage and high efficiency, (2) low-temperature fabrication processes (<200 °C) that prevent any degradation of solar grade CZ c-Si wafers, (3) excellent temperature coefficient. Figure 8 shows the progress in the efficiency of HIT solar cells (R&D and mass production). We have successfully applied our high-efficiency processes to very thin silicon wafers that are less than 100 μm thick at the R&D stage [23-35]. Figure 9 shows the I-V characteristics of HIT solar cells measured by AIST. We established a new record efficiency of 24.7% (Voc: 750 mV, Jsc: 39.4 mA/cm², F.F.: 0.832, P max: 24.7 mW/cm²) for practical-size HIT solar cells (Thickness: 98 μm, Area: 102 cm², R&D) by simultaneously improving (a) the HIT junction, (b) the optical confinement and (c) the electric contacts [36].

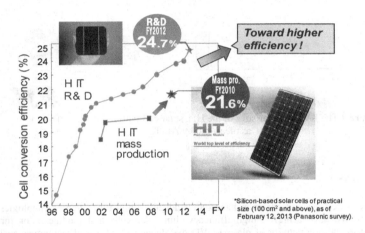

Figure 8 Progress in the efficiency of HIT solar cells (R&D and mass production).

Cell thickness: 98μ m

I-V CURVE

IEC60904-3Ed.2 101.8cm2 (total area) WXS-220S-20

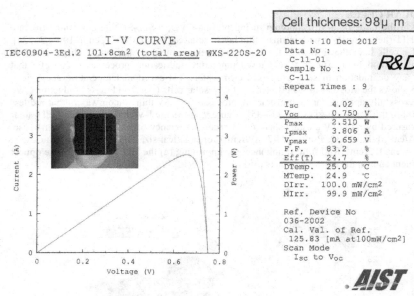

Date : 10 Dec 2012
Data No :
  C-11-01
Sample No :
  C-11
Repeat Times : 9

R&D

Isc	4.02	A
Voc	0.750	V
Pmax	2.510	W
Ipmax	3.806	A
Vpmax	0.659	V
F.F.	83.2	%
Eff(T)	24.7	%
DTemp.	25.0	℃
MTemp.	24.9	℃
DIrr.	100.0	mW/cm2
MIrr.	99.9	mW/cm2

Ref. Device No
036-2002
Cal. Val. of Ref.
  125.83 [mA at100mW/cm2]
Scan Mode
  Isc to Voc

AIST

Figure 9 The I-V characteristics of the HIT solar cell with the highest efficiency as measured by AIST.

CONCLUSIONS

We reviewed recent progress in Panasonic's advanced thin film silicon technologies for photovoltaics. We have achieved high-efficiency thin solar cells and modules. For further improvement of the productivity of plasma-CVD, we developed advanced techniques such as radical flux simulation and microstructure analysis of μc-Si, in which atomic H has proved to play an important role. An alternative process for a-Si fabrication under non-vacuum conditions, the so-called "Liquid-Si printing method" was also described. Finally, we reported a new record conversion efficiency of 24.7% for HIT solar cells using a very thin c-Si wafer.

ACKNOWLEDGMENTS

This work was supported in part by NEDO (New Energy and Industrial Technology Development Organization) under the Ministry of Economy, Trade and Industry. The authors wish to thank the ERATO SHIMODA Nano-Liquid Process Project sponsored by JST (the Japan

Science and Technology Agency) under the Ministry of Education, Culture, Sports, Science and Technology.

REFERENCES

1. Y. Hishikawa, M. Sasaki, S. Tsuge, S. Okamoto and S. Tsuda: *Mater. Res. Soc. Symp. Proc.* **297** (1993) 779.
2. Y. Hishikawa, K. Ninomiya, E. Maruyama, S. Kuroda, A. Terakawa, K. Sayama, H. Tarui, M. Sasaki, S. Tsuda and S. Nakano: *Proc. 1st World Conf. Photovoltaic Energy Conversion* (1994, Waikoloa) p. 386
3. M. Kameda, S. Sakai, M. Isomura, K. Sayama, Y. Hishikawa, S. Matsumi, H. Haku, K. Wakisaka, M. Tanaka, S. Kiyama, S. Tsuda, and S. Nakano: *Proc. 25th IEEE Photovoltaic Specialists Conf.* (1996, Washington) p. 1049.
4. A. Terakawa, M. Shima, K. Sayama, H. Tarui, H. Nishiwaki and S. Tsuda: Jpn. J. Appl. Phys. 34 (1995) 1741
5. A. Terakawa, M. Shima, T. Kinoshita, M. Isomura, M. Tanaka, S. Kiyama and S. Tsuda: *Proc. 17th European PV Solar Energy Conf.* (1997, Barcelona) 2359
6. S. Okamoto, A. Terakawa, E. Maruyama, W. Shinohara, M. Tanaka, and S. Kiyama: *Mater. Res. Soc. Symp. Proc.* **664** (2001) A11.1
7. E. Maruyama, S. Okamoto, A. Terakawa, W. Shinohara, M. Tanaka, and S. Kiyama: Sol. Energy Mater. Sol. Cells 74 (2002) 339
8. M. Matsumoto, K. Kawamoto, T. Mishima, H. Haku, M. Shima, A. Terakawa and M. Tanaka: *Proc. 4th World Conf. Photovoltaic Energy Conversion* (2006, Hawaii) 1580
9. Y. Aya, M. Matsumoto, K. Murata, S. Ogasawara, M. Nakagawa, A. Terakawa and M. Tanaka: *Technical Digest 17th Photovoltaic Science and Engineering Conf.* (2007, Fukuoka) 177
10. T. Kunii, K. Murata, M. Matsumoto, K. Kawamoto, Y. Kobayashi, Y. Aya, M. Nakagawa, A. Terakawa and M. Tanaka: *Proc. 33rd IEEE Photovoltaic Specialists Conf.* (2008, San Diego) 259
11. Y. Aya, K. Murata, H. Katayama, W. Shinohara, M. Nakagawa, A. Terakawa and M. Tanaka: *Proc. 24th European PV Solar Energy Conf.* (2009, Hamburg) 2394.
12. M. Hishida, A. Kuroda, T. Kunii, K. Murata, M. Matsumoto, Y. Aya, A. Terakawa and M. Tanaka: *Proc. 19th International Photovoltaic Science and Engineering Conf.* (2009, Jeju) 122.
13. T. Sekimoto, H. Katayama, K. Murata, M. Matsumoto, A Kitahara, M. Hishida, Y. Aya, W. Shinohara, M. Nakagawa, A. Terakawa, and M. Tanaka: *Proc. 35th IEEE Photovoltaic Specialists Conf.* (2010, Hawaii) 1147
14. H. Katayama, K. Murata, T. Kunii, M. Matsumoto, Y. Aya, W. Shinohara, A. Kitahara, M. Nakagawa, A. Terakawa and M. Tanaka, *RENEWABLE ENERGY 2010 Proc.* (2010, Yokohama) OP-14-11
15. W. Shinohara, Y. Aya, M. Hishida, N. Kitahara, M. Nakagawa, A. Terakawa and M. Tanaka, *Proc. 25th European PV Solar Energy Conf.* (2010, Valencia) 2735
16. A. Terakawa: J. Jpn. Soc. Plasma Sci. Nucl. Fusion Res. **86** (2010) 17
17. A. Terakawa, M. Hishida, S. Yata, W. Shinohara, A. Kitahara, H. Yoneda, Y. Aya, I. Yoshida, M. Iseki and M. Tanaka: *Proc. 26th European PV Solar Energy Conf.* (2011) 3BO.4.2.

18. Y. Aya, W. Shinohara, M. Matsumoto, K. Murata, T. Kunii, M. Nakagawa, A. Terakawa, M. Tanaka: Progress in Photovoltaics **20** (2012) 166

19. M. Matsumoto, Y. Aya, M. Hishida, S. Yata, W. Shinohara, I. Yoshida, D. Kanematsu, A. Terakawa, M. Iseki and M. Tanaka: *Mater. Res. Soc. Symp. Proc.* **1426** (2012) 1117

20. Y. Naruse, M. Matsumoto, T. Sekimoto, M. Hishida, Y. Aya, W. Shinohara, A. Fukushima, S. Yata, A. Terakawa, M. Iseki, and M. Tanaka: *Proc. 38th IEEE Photovoltaic Photovoltaic Specialists Conf.* (2012, Texas) 3118

21. H. Katayama, I. Yoshida, A. Terakawa, Y. Aya, M. Iseki and M. Tanaka: to be published in *Proc. 27th European PV Solar Energy Conf.* (2012)

22. S. Yata, Y. Aya, A. Terakawa, M. Iseki, M. Taguchi and M. Tanaka: to be published in *Technical Digest 22nd Photovoltaic Science and Engineering Conf.* (2012, Hangzhou)

23. L. Guo, M. Kondo, M. Fukawa, K. Saitoh and A. Matsuda: Jpn. J. Appl. Phys. 037 (1998) L1116

24. M. Kondo, M. Fukawa, L. Guo and A. Matsuda: J. Non-Cryst. Solids 266–269 (2000) 84

25. Y. Sobajima, M. Nishino, T. Fukumori, M. Kurihara, T. Higuchi, S. Nakano, T. Toyama and H. Okamoto: Sol. Energy Mater. Sol. Cells **93** (2002). 980

26. Y. Nasuno, M. Kondo and A. Matsuda, Jpn. J. Appl. Phys. **40** (2001) L303.

27. A. V. Shah., H. Schade, M. Vanecek, J. Meier, E. Vallat-Sauvain, N. Wyrsch, U. Kroll, C. Droz and J. Bailat., Prog. Photovolt: Res.**12** (2004) 113.

28. M. Python, , O. Madani, D. Dominé, F. Meillaud, E. Vallat-Sauvain1 and C. Ballif., Sol. Energy Mater Sol. Cells 93 (2009) 1714.

29. H. B. T. Li, R. H. Franken, J. K. Rath and R. E. I. Schropp, Sol. Energy Mater. Sol. Cells (2009) 338.

30. T. Shimoda, Y. Matsuki, M. Furusawa, T. Aoki, I. Yudasaka, H. Tanaka, H. Iwasawa, D. Wang, M. Miyasaka, and Y. Takeuchi: Nature 440 (2006) 783.

31. T. Masuda, N. Sotani, H. Hamada, Y. Matsuki, and T. Shimoda: Appl. Phys. Lett. 100 (2012) 253908

32. H. Murayama, T. Ohyama, I. Yoshida, A. Terakawa, T. Masuda, K. Ohdaira and T. Shimoda: to be published in Thin Solid Films

33. M. Tanaka, M. Taguchi, T. Matsuyama, T. Sawada, S. Tsuda, S. Nakano, H. Hanafusa, Y. Kuwano: Jpn. J. Appl. Phys., 31 (1992), 3518.

34. M. Taguchi, A. Terakawa, E. Maruyama, M. Tanaka: Prog. Photovoltaics Res. Appl., 13 (2005), 481

35. T. Mishima, M. Taguchi, H. Sakata, E. Maruyama: Sol. Energy Mater. Sol. Cells, 95 (2011), pp. 18–21

36. News release: http://panasonic.co.jp/corp/news/official.data/data.dir/2013/02/en130212-7/en130212-7.html

Mater. Res. Soc. Symp. Proc. Vol. 1536 © 2013 Materials Research Society
DOI: 10.1557/opl.2013.597

Hydrogen-plasma etching of thin amorphous silicon layers for heterojunction interdigitated back-contact solar cells

Stefano. N. Granata[1, 2], Twan Bearda[1], Ivan Gordon[1], Jef Poortmans[1,2], Robert Mertens[1,2]

[1]IMEC, Kapledreef 75, B-3001 Heverlee, Belgium

[2]KU Leuven ESAT Kasteelpark Arenberg 10 B-3001 Heverlee Belgium

ABSTRACT

In this study, A H_2-plasma is studied as a dry method to etch thin layers of amorphous silicon aSi:H(i) deposited on a crystalline wafer. It is found that H_2-plasma etches aSi:H(i) selectively toward silicon nitrides hard masks with an etch rate below 3nm/min. Depending on power density and temperature of the substrate during the H_2-plasma, the energy bandgap, the hydrides distribution and the void concentration of the aSi:H(i) layers are modified and the amorphous-to-crystalline transition is approached. At high temperature (>250C) and low plasma power (<20mW/cm^2), the dihydride (SiH$_2$) content increases and the bandgap widens. The etch rates stays below 0.5 nm/min. At low temperature (<150°C) and high power (>70mW/cm^2), the void concentration increases significantly and etch rates up to 3nm/min are recorded. These findings are supported by a theoretical model that indicates formation of Si-H-Si precursors in the layer during exposure to H_2-plasma. According to the experimental conditions, these precursors either diffuses and forms Si-Si strong bonds or are removed from the film, causing layer etching.

INTRODUCTION

In heterojunction back contact (HJ i-BC) solar cells, thin (<20nm) amorphous silicon (aSi:H(i)) layers of opposite polarity are deposited side by side on the rear surface of silicon wafers (cSi); this results in an interdigitated pattern of aSi:H(p) and aSi:H(n). Within this geometry experimental efficiencies up to 23.4% [1] are shown.

The process to define the aSi:H(p) - aSi:H(n) interdigitated pattern involves multiple steps. In the easiest configuration, a sequence of blanket layer deposition, photolithography and layer etching is repeated for each of the two layers [2]. Alternatively the process can include patterning of passivation layers in between fingers or hard mask definition for selective etching [3]. These processes have been inherited from high efficiency homojunction i-BC [4] and are inadequate to HJ because of differences between crystalline and amorphous layer structures. In homojonction i-BC, emitter and BSF layer are few microns thick and the doping concentration varies approximately from 1E20 cm^{-3} on the surface to 1E16 cm^{-3} nearby the space charge region [5]. Thus, the experimental conditions are tuned to afford etch rates higher than 1μm/min, and are relatively insensitive to layer doping [6]. For these reasons, solutions commonly used are HF/HNO$_3$ mixtures and concentrated TMAH at high temperature (>70°C).

In HJ i-BC, emitter and BSF thicknesses falls into the nanometer scale, and chemical doping can be considered constant and in the order of magnitude of 1E19 cm^{-3} [2]. Therefore, high etch rates are unnecessary or harmful, since the etched profile is poorly controlled.

However, lowering the etch rate is not trivial, since doping sensitiveness and surface smoothness are also affected [6].Assuming a broader perspective on the problem, dry etching can also be considered. However, fluorine or chlorine-based gases etches aSi:H(i) with rates above 100nm/min [6] and are not selective toward the underlying cSi substrate. This could result in surface damage and open circuit voltages would be negatively affected. [7].

Therefore, H_2-plasma is considered as an alternative etching method suitable for thin layers. H_2-plasma is found to remove aSi:H with an etch rate that can vary from 0.4 to 2.6 nm/min. The model underlying the etching mechanism is described and the layer modification due to the treatment is studied in terms of energy bandgap, hydride configuration and void concentration.

EXPERIMENT

All presented studies were carried following the same experiment sequence. After a first aSi:H(i) layer was deposited on the substrate, the samples underwent an H_2-plasma and were analyzed. Substrates used were n-type <100> single side optically-polished CZ wafers of a thickness of 700 μm and a resistivity above $1\Omega cm^2$. All samples were cleaned by a 2-step process employing H_2SO_4:H_2O_2 4:1 and HF:HCl:H_2O 1:1:20 prior to samples loading for the first aSi:H deposition. Before H_2-plasma samples underwent a HF dip in HF:HCl:H_2O 1:1:20. All aSi:H(i), SiN_x deposition and H_2-plasma were performed in an Oxford Instrument Plasmalab System 100 PECVD reactor, with the following conditions. The aSi:H(i) layer of a thickness of 25nm was deposited with a power density of 45 mw/cm^2, temperature of 200°C, pressure of 1.7 Torr and hydrogen dilution of 1:3. The SiN_x layer of a thickness of 110±5 nm was deposited with a power density of 100mW/cm^2, temperature of 200 C, pressure of 1.5 Torr and N_2O:SiH_4 ratio of 2:5. The H_2- plasma etching condition varied in the range of: 18-100 mW/cm^2 , 70-350 C, 1-2 Torr and 100-200 sccm. These parameters were combined through a design of experiments to cover the full experimental window. In order to verify etching reproducibility and calculate the etch rate, at least four samples underwent the same H_2 plasma at different times in the range 30-600 seconds.

All samples were measured by spectral ellipsometry and thicknesses and structural characteristics of layers were fitted with an effective medium approximation [8], combining a Tauc-Lorentz oscillator with voids fraction of 20%.The absorption of hydrides (SiH_x, x=1,2) in the film were measured by FTIR in Attenuated Total Reflectance mode (ATR). The ration of SiH_2 over the total number of hydrides was obtained comparing the absorption peaks at 2090 cm[-1] (SiH_2) and 2000cm[-1] (SiH) and the microstructural factor R [9] was calculated (eq. 1).

$$R = \frac{I_{2090}}{I_{2000} + I_{2090}}$$
(1)

DISCUSSION

Figure 1 shows the thickness decreases of aSi:H(i) and SiNx as a function of the exposure time to H_2-plasma.

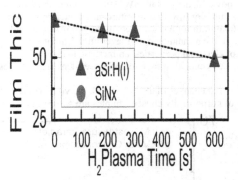

Figure 1 aSi:H(i) and SiNx thickness variation as a function of H_2 plasma time. The H_2 plasma conditions were: 48mW/cm2, 250C, 200sccm, 2Torr.

In this case, only one set of experimental conditions were chosen for the H_2-plasma, in order to prove etching selectivity. From the graph it is evident how only the aSi:H(i) is etched, while changes in SiNx thickness remains within the error bar. Figure.2(left) shows the etch rate of the aSi:H(i) layer as a function of temperature and power density of the H_2-plasma.

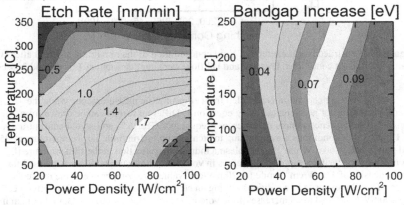

Figure 2 Etch rate (left) and Bandgap (right) Increase of aSi:H(i) as a function of Temperature and Power Density. Pressure and H_2 flow were kept constant at 1.5Torr and 200sccm

This results are obtained by polynomial interpolation of experimental etch rates under different conditions. In the experimental window considered, the variation of the etch rate as a function of the H_2 flow and pressure is not significant and hence not shown.

The experimental etch rate is influenced from the combined effect of temperature and power density. At low power density (<30 mW/cm^2) and high temperature (>250C) the etch rate recorded is below 1nm/min. For lower temperature and higher power density the etch rate increases and attain 2.6 nm/min for a power density of 100 mW/cm^2 and a temperature of 50°C.The modification in the remaining aSi:H layer after 5minutes of H_2–plasma are also investigated. In terms of electrical properties, an increase in energy bandgap is described in Fig.2(right). This increase occurs as a function of power density, regardless of the etching temperature.

In terms of structural changes of aSi:H network the void concentration and the microstructure factors of purposefully chosen samples are shown in Figure. 3.

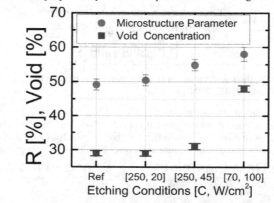

Figure 3 Microstructure parameter and Void concentration of the remaining aSI:H(i) after H2-plasma at different conditions. Pressure and H2 flow were kept constant at 1.5Torr and 200sccm

The void concentration is modeled from ellipsometric measurements, while the microstructure factor is calculated using equation 1. Different results can be drawn from Figure.3. The reference sample indicates an R equal to 49.2±1.6 % and a void concentration of 28.1±0.8%. For conditions corresponding to etch rate up to 1nm/min, i.e. Temperature of 250°C and power below 60 mW/cm^2, the H_2-plasma induces modification in the film structure if enough power density is supplied. Increases in void concentrations and R are witnessed for power density of 45mW/cm^2, while for 20mW/cm^2 values are comparable to the reference.

The film exposed to conditions giving an etch rate of approximately 2 nm/min, i.e. 70°C and 100mW/cm^2, lead to an increase of the void concentration by 48.0±0.8 %, higher than the reference. The variation of SiH_2 fells in the error bar, and cannot be considered significant.

The model proposed by Boland et al. [10] is considered to explain the experimental findings. During aSi:H film exposure to H_2-plasma, hydrogen atoms penetrate the layer. Surface and subsurface reactions of addition, abstraction, insertion and etching can take place [11] and the hydrogen content in the material increases. Usually, this is accompanied by an increase in the energy bandgap [12]. In case of insertion, the hydrogen breaks strained Si-Si bonds, and forms Si-H-Si precursors. Precursor can then diffuse through the aSi:H layer until they reach a favorable binding site. At this moment, hydrogen is released and a strong Si-Si bonds is formed. Subsequently to this modification the layer approaches the amorphous-to-crystalline transitions [13]. However, a lack of binding site leads to precursor removal and hence, film etching.

The changes in macroscopic properties witnessed in Figure. 2 and 3 indicates that after treatment the film approaches the amorphous-crystalline transition. Therefore, the model by Boland et al. is valid in the experimental window considered and it is used to interpret the evolution of the H_2 etching rate as a function of temperature and power density.

The incorporation of hydrogen in aSi:H(i) is only possible if plasma species have enough kinetic energy to penetrate the film [14]. This kinetic energy is the energy attributed to radicals and ions after plasma dissociation and ionization. At greater power densities, the average kinetic energy of the plasma discharge is higher, therefore a higher number of hydrogen species can penetrate the film. For this reason, the increase in energy bandgap related to the higher hydrogen content of the film is only proportional to the power density. This is not the case for the etching rate, the void concentration and the microstructure factor, where also the temperature seems to play a role. Indeed, in order to induce structural modifications of the aSi:H(i) network, not only has the hydrogen to penetrate the film, but the Si-H-Si precursor needs to diffuse through the aSi:H(i) layer to reach a suitable binding site. Being the diffusion a thermal activated process [15], the temperature of the film during the etching treatment helps to regulate the diffusion of the precursor in the film. Therefore, at higher temperature, most of the diffusing species reach a suitable binding site, with formation of Si-Si bonds. The film reorganizes and approaches the amorphous-to-crystalline transition. At low temperature, the probability to find a suitable binding site is lower. The precursors are removed from the surface and the film is etched. Layer reorganization does not occur and the void fraction increases considerably.

CONCLUSIONS

The H_2-plasma etching of a thin amorphous intrinsic silicon layer for heterojunction back contact solar cells is studied. Etch rates up to 2.6nm/min are achieved and structural modifications in the remaining aSi:H layer are investigated. Both structural modification and etching takes place during an H_2-plasma and the predominance of one mechanism upon the other is regulated by temperature and power density. At power density below 20mW/cm^2, no significant modifications of the film occur. If enough power density is supplied, modifications in the amorphous film are witnessed: at high temperature (>250°C) the film tends towards crystalline transition and at low temperature the film is etched. This change in regime is mainly due to the temperature dependent diffusion of the hydrogen incorporated in the a-Si:H network.

As such, hydrogen plasma can be used to remove thin layer of amorphous silicon from a crystalline substrate, since it respects criteria of low etch rate and selectivity toward hard masks. However, the integration of this step in a solar cell process flow would be possible only if the amorphous/silicon interface is damage free after H_2 etching of the full layer. Surface damage and interface defects induced by H_2–plasma will be object for future investigations.

REFERENCES
1. De Wolf Stefaan, Descoeudres Antoine, Holman Zachary C., and Ballif Christophe, Green **2**, 7 (2012)
2. M. Lu, U. Das, S. Bowden, S. Hegedus, and R. Birkmire, Progress in Photovoltaics: Research and Applications **19**, 326–338 (2011).
3. N. Mingirulli, J. Haschke, R. Gogolin, R. Ferré, T.F. Schulze, J. Düsterhöft, N.-P. Harder, L. Korte, R. Brendel, and B. Rech, Physica Status Solidi (RRL) – Rapid Research Letters **5**, 159–161 (2011).
4. P.J. Verlinden, M. Aleman, N. Posthuma, J. Fernandez, B. Pawlak, J. Robbelein, M. Debucquoy, K. Van Wichelen, and J. Poortmans, Solar Energy Materials and Solar Cells **106**, 37 (2012).
5. M. Aleman, J. Das, T. Janssens, B. Pawlak, N. Posthuma, J. Robbelein, S. Singh, K. Baert, J. Poortmans, J. Fernandez, K. Yoshikawa, and P.J. Verlinden, Energy Procedia **27**, 638 (2012).
6. G.T.A. Kovacs, N.I. Maluf, and K.E. Petersen, Proceedings of the IEEE **86**, 1536 (1998).
7. M. Tucci, E. Salurso, F. Roca, and F. Palma, Thin Solid Films **403-404**, 307 (2002).
8. D.E. Aspnes, J.B. Theeten, and F. Hottier, Phys. Rev. B **20**, 3292 (1979).
9. A.H. Mahan, P. Menna, and R. Tsu, Applied Physics Letters **51**, 1167 (1987).
10. J.J. Boland and G.N. Parsons, Science **256**, 1304 (1992).
11. C.-M. Chiang, S.M. Gates, S.S. Lee, M. Kong, and S.F. Bent, J. Phys. Chem. B **101**, 9537 (1997).
12. T.F. Schulze, L. Korte, F. Ruske, and B. Rech, Phys. Rev. B **83**, 165314 (2011).
13. A. Descoeudres, L. Barraud, S. De Wolf, B. Strahm, D. Lachenal, C. Guérin, Z.C. Holman, F. Zicarelli, B. Demaurex, J. Seif, J. Holovsky, and C. Ballif, Applied Physics Letters **99**, 123506 (2011).
14. B.N. Chapman, *Glow Discharge Processes: Sputtering and Plasma Etching* (Wiley, 1980).
15. C.G. Van de Walle and R.A. Street, Phys. Rev. B **51**, 10615–10618 (1995).

Mater. Res. Soc. Symp. Proc. Vol. 1536 © 2013 Materials Research Society
DOI: 10.1557/opl.2013.749

Employing μc-SiOX:H as n-Type Layer and Back TCO Replacement for High-Efficiency a-Si:H/μc-Si:H Tandem Solar Cells

S.W. Liang, C.H. Hsu, Y.W. Tseng, Y.P. Lin and C.C. Tsai
Department of Photonics, National Chiao Tung University, 1001 University Road, Hsinchu, Taiwan

ABSTRACT

The n-type hydrogenated microcrystalline silicon oxide (μc-SiO$_X$:H(n)) films with different stoichiometry have been successfully prepared by varying the CO_2-to-SiH_4 flow ratio in the PECVD system. By using the μc-SiO$_X$:H(n) as a replacement for μc-Si:H(n) and ITO, the conversion efficiency of μc-Si:H single-junction and a-Si:H/μc-Si:H tandem cells were improved to 6.35% and 10.15%, respectively. The major improvement of the short circuit current density (J_{SC}) and these cell efficiencies were originated from the increased optical absorption, which was confirmed by the quantum efficiency measurement showing increased response in the long-wavelength region. Moreover, the all PECVD process except the metal contact simplified the fabrication and might benefit the industrial production.

INTRODUCTION

In the applications of silicon thin-film solar cells, the most popular stacked cell configuration is the a-Si:H/μc-Si:H tandem cell which consists of a higher bandgap amorphous silicon (a-Si:H) top absorber and a narrower bandgap microcrystalline silicon (μc-Si:H) bottom absorber [1]. In order to reduce the light-induced degradation in a-Si:H and the long deposition time of μc-Si:H, the thickness of a-Si:H and μc-Si:H absorbers have to be kept reasonably thin while provide sufficient photocurrent. The well-controlled light-management for using thinner absorbers is then essential for high performance a-Si:H/μc-Si:H tandem cells. It permits to achieve longer light paths in the absorber, allowing for the use of thinner active layers [2]. For achieving adequate top cell current, the optical absorption in top cell can be increased by using an intermediate reflecting layer between the top and the bottom cells [3-4]. To obtain sufficient bottom cell current, an effective approach is the use of a highly reflective back reflector containing a TCO and a metal contact [5-6]. The back TCO is employed to improve the refractive index matching between silicon thin films and the metal to allow the un-absorbed photons to be reflected back into the absorber. Hegedus et al. have also shown μc-Si:H(n)/ITO/Ag back reflecting (BR) structure had better electrical contact than a-Si:H(n)/ITO/Ag BR structure in a-Si:H single-junction solar cells [7]. Moreover, the insertion of TCO was proposed to reduce the absorption losses due to the decreased excitation of texture mediated surface plasmonic resonances at the silver metal contact [8]. However, an ex-situ sputtering step for TCO is needed. Without the ex-situ sputtering step for back TCO, the capability of n-type hydrogenated microcrystalline silicon oxide (μc-SiO$_X$:H) as an alternative to n-layer and back TCO in a-Si:H solar cells was presented [9-10].

The electrical and optical properties of μc-SiO$_X$:H films significantly depends on the film content and crystallization. The characteristics of μc-SiO$_X$:H can be controlled over a wide range by in-situ varying the deposition conditions in PECVD, for example, studies have reported that the oxygen concentration can be adjusted by CO_2-to-SiH_4 flow ratio. The bandgap (E_{04}) can be controlled from below 2 eV up to a value of 2.9 eV, and the refractive index (n) can be controlled from approximately 3.8 (μc-Si:H) down to approximately 1.5 which roughly corresponds to n of silicon dioxide [11-12]. Depending on the requirement of the devices, μc-SiO_X:H film can be prepared by considering the trade-off between electrical and optical properties. In this work, we used μc-SiO_X:H film as a replacement for n-layer and back TCO in μc-Si:H single-junction and bottom cells.

EXPERIMENTAL DETAILS

The doped and undoped silicon-based thin films were deposited by a 27.12MHz single-chamber PECVD system. The a-Si:H/μc-Si:H tandem cells were made on textured SnO_2:F TCO glasses with a superstrate configuration. The n-type μc-SiO_X:H films were deposited by introducing PH_3 and CO_2 with highly H_2-diluted SiH_4. The film content and crystalline volume fraction of μc-SiO_X:H films were examined by a X-ray photoelectron spectroscopy (XPS) and a Raman spectroscopy (λ=488 nm), respectively. Electrical measurements were carried out to investigate the conductivity. The bandgap and the refractive index of μc-SiO_X:H layer were obtained by optical measurements. An AM1.5G illuminated J-V measurement system and a quantum efficiency (QE) instrument were used to measure the cell performance. The cell area was defined by the metal electrode which is $0.25 cm^2$.

RESULTS AND DISCUSSION

Films with different stoichiometry have been deposited by varying the CO_2-to-SiH_4 flow ratio (R_{CO2}) in the range of 0.53–3.53. Figure 1 shows the effect of R_{CO2} on oxygen content (C_O), crystalline volume fraction (X_C) and conductivity (σ) in μc-SiO_X:H(n) films. Electrical and optical measurements were carried out on 100-nm-thick μc-SiO_X:H(n) layers. The hydrogen dilution ratio (R_{H2}, defined as the ratio of hydrogen flow rate to the total gas flow rate) was kept at 98.5% with the PH_3 flow rate of 5 sccm. As the R_{CO2} increased from 0.53 to 3.53, the oxygen content increased from 3.5 to 37.3 at.%. The introduced CO_2 was suggested to react with hydrogen and forms OH radical [13]. The OH radical bonds to the film surface and thus the O is incorporated into the film. However, the crystalline volume fraction and the conductivity both decreased as the R_{CO2} increased. The X_C decreased from 57% to 16%, and the conductivity decreased from 5.31 to 3.41×10^{-9} S/cm. The decrease in X_C was attributed to the increase in oxygen content, which quenched the crystallization in μc-SiO_X:H(n) films. The increase in oxygen content also increased the hydrogenated amorphous silicon oxide (a-SiO_X:H) phase which led to lower refractive index and higher bandgap [13]. As a result, the fewer amount of crystallization phase in μc-SiO_X:H(n) films led to the decrease in conductivity. Meanwhile, the n-type hydrogenated microcrystalline silicon (μc-Si:H) contributed to the crystalline volume fraction as well as the sufficient electrical conductivity.

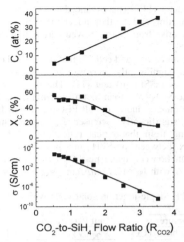

Fig. 1. Effect of CO_2 to SiH_4 flow ratio (R_{CO2}) on oxygen content (C_O), crystalline volume fraction (X_C) and conductivity (σ)

N-type hydrogenated microcrystalline silicon oxide ($\mu c\text{-}SiO_X:H(n)$) has a larger bandgap and a lower refractive index, with acceptable conductivity compared to n-type microcrystalline silicon ($\mu c\text{-}Si:H(n)$) [11]. To investigate the effect of $\mu c\text{-}SiO_X:H(n)$ layers in the solar cells, we employed $\mu c\text{-}SiO_X:H(n)$ to replace $\mu c\text{-}Si:H(n)$ and ITO in a $\mu c\text{-}Si:H$ single-junction solar cell. The cell structure is shown in Fig. 2.

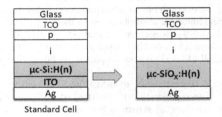

Fig. 2. Schematic diagram of the cell structure showing the replacement of $\mu c\text{-}Si:H(n)$ and ITO by $\mu c\text{-}SiO_X:H(n)$ in $\mu c\text{-}Si:H$ single-junction solar cells

Table 1 shows the performance of $\mu c\text{-}Si:H$ single-junction solar cells with different back reflecting structure. The thickness of $\mu c\text{-}Si:H$ absorber was 1200 nm, and the $\mu c\text{-}SiO_X:H(n)$ films were deposited with gas flow ratios of $R_{CO2}=0.55$ and $PH_3/SiH_4=0.45\%$. Compared to cell #1, cell #2 had an enhancement of short circuit current density (J_{SC}) from 15.16 to 15.94 mA/cm^2

and slightly increased in fill factor (FF). The enhanced J_{SC} can be attributed to the refractive index matching between silicon films and Ag that induced the optical light path in the absorber. The slightly increased FF may arise from the improved carrier transport between silicon films and Ag by ITO.

Moreover, it is observed from cell #2 and cell #3, that the FF and open circuit voltage (V_{OC}) slightly increased with the J_{SC} significantly increased from 15.94 to 17.74 mA/cm^2 by the replacement of μc-SiO$_X$:H(n) for μc-Si:H(n) and ITO. The increased V_{OC} may be due to the better electrical properties of μc-SiO$_X$:H(n) than that of μc-Si:H(n), and the increased FF can be ascribed to the better interface quality due to the all PECVD in-situ fabrication by eliminating the sputtering ITO step. The significantly increased J_{SC} can be attributed to more optical absorption in absorbers, arising from the optical reflection at the i/n interface due to the difference of refractive index between μc-Si:H and μc-SiO$_X$:H(n) films. The conversion efficiency of μc-Si:H single-junction cell using μc-SiO$_X$:H(n) as a replacement for μc-Si:H(n) and ITO was improved to 6.35%, with J_{SC}=17.73 mA/cm^2, V_{OC}=0.52 V and FF=69.51%.

Table 1. Performance of μc-Si:H single-junction solar cells with different back reflecting (BR) structure

Back Reflecting Structure		Cell Performance			
# n-layer	back TCO	V_{OC} (V)	J_{SC} (mA/cm²)	FF (%)	η (%)
1 μc-Si:H(n)	-	0.51	15.16	67.76	5.24
2 μc-Si:H(n)	ITO	0.51	15.94	68.89	5.60
3 μc-SiO$_X$:H(n)	-	0.52	17.74	69.51	6.35

As shown in Fig. 3, the significant enhancement of J_{SC} in μc-Si:H single-junction cells was investigated by a quantum efficiency (QE) instrument. Compared to μc-Si:H(n)/Ag and μc-Si:H(n)/ITO/Ag BR structure, the use of μc-SiO$_X$:H(n)/Ag BR structure showed a significant increase in the long wavelength region of 600-1100 nm. The reason should be due to the better electrical properties of μc-SiO$_X$:H(n) to facilitate carrier collection, and the refractive-index difference enhanced the long-wavelength optical absorption in the absorber. Moreover, the larger bandgap of μc-SiO$_X$:H(n) can reduce the absorption loss in the n-type layer compared to the μc-Si:H(n) films. As a result, the QE J_{SC} was enhanced from 15.14 to 17.35 mA/cm^2 by using μc-SiO$_X$:H(n)/Ag as the replacement for μc-Si:H(n)/Ag or μc-Si:H(n)/ITO back reflecting structure. Except the metal contact, the all PECVD process simplified the fabrication and might benefit the industrial production.

Figure 4 shows the J-V characteristics of a-Si:H/μc-Si:H tandem solar cell using μc-SiO$_X$:H(n)/Ag BR structure. The superstrate cell consisted of a 220-nm-thick a-Si:H absorber and 1200-nm-thick μc-Si:H absorbers, as shown in the inset of Fig. 4. The μc-SiO$_X$:H(n) was employed in the μc-Si:H bottom cell as a replacement for μc-Si:H(n) and ITO. With appropriate optical and electrical properties of μc-SiO$_X$:H(n), the J_{SC} in the bottom cell as high as 10.40 mA/cm^2 was obtained, which was confirmed by the quantum efficiency measurement. The conversion efficiency of the a-Si:H/μc-Si:H tandem cell using μc-SiO$_X$:H(n)/Ag BR structure in the bottom cell was improved to 10.15%, with J_{SC}=10.43 mA/cm^2, V_{OC}=1.29 V and FF=75.41%.

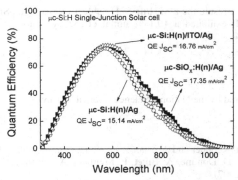

Fig. 3. Quantum efficiency of μc-Si:H single-junction cells with different back reflecting structures

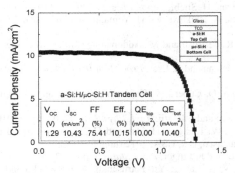

Fig. 4. J-V characteristics of a-Si:H/μc-Si:H tandem cells using the μc-SiOₓ:H(n)/Ag back reflecting structures in μc-Si:H bottom cells

CONCLUSIONS

Phosphorous doped hydrogenated microcrystalline silicon oxide (μc-SiOₓ:H(n)) films with different stoichiometry have been successfully prepared by varying the CO_2-to-SiH_4 flow ratio in the PECVD system. The oxygen content can be adjusted from 3.5 to 37.3 at.%, and the corresponding conductivity decreased from 5.31 to 3.41×10^{-9} S/cm. By using the μc-SiOₓ:H(n) as a replacement for μc-Si:H(n) and ITO, the conversion efficiency of μc-Si:H single-junction cell was improved to 6.35%, with J_{SC}=17.73 mA/cm², V_{OC}=0.52 V and FF=69.51%. The major improvement of these cell efficiencies were originated from the increased optical absorption, which was confirmed by the quantum efficiency measurement showing increased response in the long-wavelength region. As a result, the a-Si:H/μc-Si:H tandem cell using μc-SiO$_X$:H(n)/Ag BR structure in the bottom cell exhibited a 10.15% initial efficiency, with J_{SC}=10.43 mA/cm^2, V_{OC}=1.29 V and FF=75.41%.

ACKNOWLEDGMENTS

This work was sponsored by the National Science Council in Taiwan.

REFERENCES

1. J. Meier, R. Flückiger, H. Keppner and A. Shah, *Appl. Phys. Lett.* **65**, 860 (1994).
2. P. Buehlmann, J. Bailat, D. Domine, A. Billet, F. Meillaud, A. Feltrin and C. Ballif, *Appl. Phys. Lett.* **91**, 143505 (2007).
3. D. Fischer, S. Dubail, J.A. Anna Selvan, N. Pellaton Vaucher, R. Platz, Ch. Hof, U. Kroll, J. Meier, P. Torres, H. Keppner, N. Wyrsch, M. Goetz, A. Shah and K.-D. Ufert, *Proc. 25th IEEE PVSC*, Washington D.C., USA, 1053 (1996).
4. P. Obermeyer, C. Haase and H. Stiebig, *Appl. Phys. Lett.* **92**, 181102 (2008).
5. K. Hayashi, K. Masataka, A. Ishikawa and H. Yamaguchi, *Proc. 1st IEEE WCPEC*, Hawaii, USA, 674 (1994).
6. E. Terzini, A. Rubino, R. de Rosa and M. Addonzonio, *Mater. Res. Soc. Symp. Proc.* **377**, 681 (1995).
7. S.S. Hegedus, W.A. Buchanan and E. Eser, *Proc. 26th IEEE PVSC*, Anaheim, CA, USA, 603 (1997).
8. F.J. Haug, T. Söderström, O. Cubero, V. Terrazzoni-Daudrix and C. Ballif, *J. Appl. Phys.* **104**, 064509 (2008).
9. P.D. Veneri, L.V. Mercaldo and I. Usatii, *Appl. Phys. Lett.* **97**, 023512 (2010).
10. Y.P. Lin, Y.W. Tseng, S.W. Liang, C.H. Hsu and C.C. Tsai, presented at the 2012 MRS Spring Meeting, San Francisco, CA, USA, 2012 (unpublished).
11. T. Grundler, A. Lambertz and F. Finger, *Phys. Status Solidi C* **7**, 1085 (2010).
12. V. Smirnov, O. Astakhov, R. Carius, Yu. Petrusenko, V. Borysenko and F. Finger, *Jpn. J. Appl. Phys.* **51**, 022301 (2012).
13. D. Das, S.M. Iftiquar and A.K. Barua, *J. Non-Cryst. Solids* **210**, 148 (1997).

Mater. Res. Soc. Symp. Proc. Vol. 1536 © 2013 Materials Research Society
DOI: 10.1557/opl.2013.750

A Vertical PN Junction Utilizing the Impurity Photovoltaic Effect for the Enhancement of Ultra-thin Film Silicon Solar Cells

D. J. Paez[1], E. Huante-Ceron[1] and A. P. Knights[1]
[1]McMaster University, Department of Engineering Physics, 1280 Main Street West, Hamilton, ON L8S 4L7, Canada.

ABSTRACT

We report a preliminary study on the influence of indium doping on ultra-thin film silicon solar cells. The design of the cell reported here is such that it should elucidate the impact of the indium dopant, which is concentrated in the thin film. Indium, a deep level in silicon (0.157 eV above the valence band), acts as a p-type dopant *and* a sensitizer. Absorption through sub-bandgap transitions is expected based on the previously reported Impurity PhotoVoltaic (IPV) Effect [1]. It is proposed that the implementation of a novel vertical PN junction configuration together with the IPV effect enhances the efficiency of ultra-thin solar cells. The most efficient cell fabricated to date, in our research group, has a conversion efficiency of 4.3 % (active silicon thickness of 2.5 µm), a short-circuit current density of 14.9 mA/cm^2 and an open-circuit voltage of 0.51 V under 1 sun illumination. The cell has not been optimized with any type of light trapping technique and 11.24 % of the cell surface is covered by the metal contacts. Numerical simulation indicates that for the geometry used, the maximum efficiency that may be expected is 9.8 % (compared to the 4.3 % measured).

INTRODUCTION

There has always been a trade-off between efficiency and cost in the solar cell industry. The ultimate goal is to achieve a low-cost, high-efficiency solar cell that could revolutionize photovoltaics.

The strategy of our development program is based on a combination of generation II and III solar cells [2]. The idea is to minimize the cost of the cell by using less material, while remaining compatible with a CMOS process, and improving the collection efficiency using a novel PN junction geometry. In this work, we achieve the latter by arranging the PN junction periodically with a separation distance smaller than the minority carrier diffusion length. The structure and orientation of the PN junction is unconventional compared to the current commercial cells in which the PN junction is perpendicularly oriented to the incident light. This work is also designed as a first step to investigate the IPV effect [1]. The cells are fabricated on a 2.5 µm c-Si thin film in order to elucidate the advantages of using indium as a deep level dopant suitable for exploitation of the IPV effect. While the absorption efficiency in this geometry is relatively poor compared to thick crystalline silicon cells, the indium dopant is concentrated in the thin film region and thus we attempt to increase the 'signal-to-noise' ratio (i.e. we wish to maximize the impact of the indium doping) of our experimental structure.

THEORY

A solar cell has diode like characteristics when there is no illumination [3]. Therefore, it can be considered as a large PN junction. The junction is formed when the p-type and n-type semiconductor regions are coincident. At equilibrium, a built-in electric field is generated sweeping away the carriers around the junction creating the depletion region. The built-in electric field is considered one of the most important characteristics of the solar cell. Such fields allow carrier separation before recombination occurs.

When the cell is illuminated, photocurrent is generated as the solar cell absorbs incident photons with energies greater than the material's bandgap (E_g). An efficient cell converts the majority of absorbed photons into electron-hole (E-H) pairs. The E-H pairs are unstable and need to be collected before recombination occurs in order for them to participate in the generation of photocurrent. This collection process can be altered depending on the geometry of the electrical contacts. In general contacts should be engineered to extract carriers in a time short compared to their lifetime.

The Impurity Photovoltaic (IPV) Effect

The IPV effect was proposed as a method to enhance the absorption of photons whose energies are less than the semiconductor bandgap [1]. Utilization of the IPV effect presumes that photons with energies less than E_g are absorbed in a two-step excitation process through an impurity level (often described as a deep or trapping level) introduced intentionally within the silicon bandgap. The first step assumes the absorption of photons with energy (E_1) higher than the energy difference between a trapping level (E_T) in the bandgap and the valence band (E_V). The second step is induced by the absorption of photons whose energies (E_2) are less than E_g, but higher than the energy difference between the conduction band (E_C) and the trapping level [1].

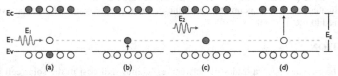

Figure 1: Impurity Photovoltaic (IPV) effect. (a) Photons with energies (E_1) less than E_g, but greater than the energy difference between E_T and E_V are absorbed promoting an electron to an available trapping level (b). (c) The second step assumes the absorption of photons with energies (E_2) less than E_g, but greater than the energy difference between E_c and E_T in order to promote the electron to the conduction band (d).

As demonstrated by Keevers *et al.* [1], indium could improve the absolute efficiency of a silicon solar cell by as much as 2% by introducing a deep-level positioned 0.157 eV above E_v. The improvement in efficiency is demonstrated to be by long-wavelength absorption. Although it is anticipated that indium would reduce the open-circuit voltage of the cell, this is greatly compensated by the improvement in photocurrent [1].

Vertical PN junctions and Periodicity

E-H pairs generated at the depletion region of a silicon junction are those most likely to be collected. On the other hand, E-H pairs generated within one diffusion length from the depletion region are collected with less efficiency.

By having a vertical PN junction oriented parallel to the surface normal, it is intended to increase the generation and collection of the photogenerated E-H pairs by increasing the depletion region volume. This is achieved by extending the length of the PN junction along the direction of propagation of the incident light instead of perpendicular to it as it is normally done, and by the placement of successive PN junctions along the surface of the cell. It is known that carrier collection decreases exponentially as one moves away from the depletion region. Therefore, by placing multiple PN junctions in close proximity, as a periodic array of PN junctions, E-H pairs generated at the *p*-type and *n*-type regions would only be required to travel distances on the order of the diffusion length before interacting with a PN junction. The collection probability is expected to be deemed "efficient" if the separation between depletion regions is no more than two diffusion lengths. This should ensure that E-H pairs generated within one diffusion length from a depletion region are collected and can participate in the generation of photocurrent.

DEVICE STRUCTURE

The design of the current solar cell is based on a periodic array of PN junctions, previously used in our research group for optical modulation in silicon waveguides [4]. The proposed structure is shown in Figure 2. The design uses silicon on insulator (SOI) as the base structure. The thickness of the active layer is 2.5 μm (c-Si) with a 1 μm buried oxide (BOX). The area of the solar cell is 0.5 cm x 0.5 cm including the metal contacts. The fabrication is made with standard contact photolithography and the variation in the doping concentration is produced by ion implantation. The implantation process is the most suitable to control the doping concentration and profiles in structures such as that shown in Figure 2, although it does add additional fabrication cost compared to diffusion based doping technologies. Although the trade-off between fabrication cost and improved efficiency is an important consideration for any device, it is not quantified in this preliminary work.

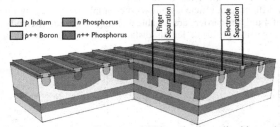

Figure 2: Proposed solar cell design. SOI based solar cell with comb fingers.

The separation between aluminum electrodes (electrode separation) is varied from 20-70 μm. Also, in order to obtain the optimum periodicity, the separation between PN junctions (separation between phosphorus 'fingers') is varied from 1, 2, 3 and 4 μm. Five device groups (each with 9 solar cells) were fabricated using indium as a dopant to form the *p*-region (the *p*++ region always being formed via boron doping). The ratio of the n dopant (phosphorus) to the p dopant (indium) was maintained at 3, while the absolute concentrations of n and p doping were varied. The *n*++ and *p*++ doping concentrations were kept constant for all samples to ensure ohmic contacts. The solar cells were characterized using a solar simulator, which was calibrated using a reference cell to 1 sun. Values for the short-circuit current (I_{sc}), open-circuit voltage (V_{oc}), current at maximum power (I_{MP}), voltage at maximum power (V_{MP}), fill factor (FF), and conversion efficiency (η) were used to determine the efficiency of various configurations.

RESULTS AND DISCUSSION

Figure 3 shows an SEM surface image of the fabricated solar cell. The difference in oxidation rates due to the different doping concentrations allows observation of the various components of the cell.

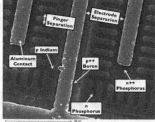

Figure 3: SEM image of final solar cell showing the well-defined finger interdigitation of the PN junctions.

The results of the data analysis are shown in figure 4. There is considerable variation in the cell efficiencies, which reflect the various doping and electrode configurations; however one consistent trend is that electrode separations smaller than 35 μm decrease the conversion efficiency. This result is consistent with an increase in surface area covered with aluminum for smaller electrode separation. A further result elucidated by a detailed ANOVA analysis is that there is no significant difference in efficiency among the samples with varying finger separation. Even at a separation of 4 μm, the carriers are capable of being collected efficiently. This suggests that the two-minority carrier diffusion length distance is still not reached at 4 μm comb separation.

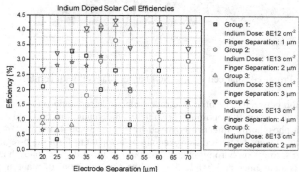

Figure 4: Experimental results for the five groups of solar cells implanted with indium.

The most efficient cell fabricated shows a conversion efficiency of 4.33 % for an active thickness of 2.5 μm. Previously reported thin film solar cells have demonstrated a theoretical limiting efficiency of 9.07 % for a 2 μm thick c-Si solar cell with optimized anti-reflective coating [5] and a maximum experimental efficiency of 15.4 % for cells with a thickness of 25.5 μm [6]. These previously reported results indicate in particular the potential effects of light trapping techniques on the current cells. At this time and due to the fact that our cell efficiencies are only 50% of those reported in [5] we are unable to comment with certainty and in a quantitative manner on the impact in efficiency attributed to the IPV effect. To address the impact of the IPV effect we have instigated the fabrication of cells which use boron in place of indium as a dopant for the p-region, although fabrication of these comparative cells was not complete at the time of writing.

Numerical simulations were developed in order to study the effect of an optimized anti-reflective (AR) coating and thickness of the BOX on the photocurrent conversion efficiency, and to place an upper-bound on the efficiency one might expect from our experimental cell. The numerical simulations are based on the model proposed by J. Prentice [7]. The model assumes sunlight is incident on the cell, which is covered with an AR coating of thickness d_1. The second layer (thin silicon) is of thickness d_2 (which in our design is of 2.5 μm). The third layer of thickness d_3 is the BOX, whose main function is to reflect most of the light that was not absorbed by the silicon layer- it thus behaves as a reflecting back layer. This model is shown in Figure 5.

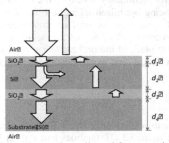

Figure 5: Ultra-thin silicon solar cell used for numerical model.

In the simulation the thickness of the silicon layer is kept fixed at 2.5 μm and the thickness of the AR coating, d_1, is varied. The photocurrent density is calculated for five different thicknesses of the buried oxide, d_3, assuming that all absorbed light is converted to E-H pairs. The photocurrent density values obtained are plotted in Figure 6a. A maximum is identified for an anti-reflecting coating thickness, d_1, of 89 nm. Similar trends were observed for different thicknesses of the buried oxide, with an absolute photocurrent maximum observed for a thickness of the buried oxide of 125 nm. Figure 6b shows the photocurrent density for an optimal AR coating of 89 nm and a silicon layer of 2.5 μm for different thickness of the buried oxide. The maximum limiting efficiency for a solar cell with active thickness of 2.5 μm was calculated from the model to be 12 % with an associated current density (J_{sc}) of 15.6 mA/cm^2. Whereas, the simulation using parameters coincident with our experimental thin cells reported above indicate a limiting efficiency of 9.8 %. The less than maximum carrier extraction for our experimental cell is likely a result of carrier recombination at the BOX/Si interface and other expected recombination centers such as the surface oxide/Si interface and residual process defects.

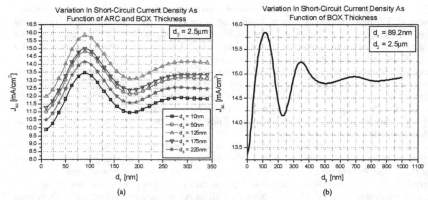

Figure 6: (a) Photocurrent density as a function of an anti-reflective coating and buried oxide thickness for a silicon layer of 2.5 μm and (b) optimized thicknesses for this silicon layer.

The current work shows preliminary results in the development of a thin film solar cell. The cell design is currently being optimized taking into account the results from experimental data analysis and numerical simulation as well as implementing other techniques that are expected to improve the conversion efficiency of the device (such as anti-reflective coating). Solar cells containing boron (in place of indium) are to be fabricated as a direct comparison and it is hoped that this will provide the first experimental determination of the impact of the IPV effect.

ACKNOWLEDGMENTS

The authors would like to acknowledge Dr. R. R. LaPierre and Dr. J. S. Preston for useful discussions. Special gratitude to the CEDT facilities and staff at McMaster University for their constant advice and support. D. J. Paez and E. Huante-Ceron would like to acknowledge the NSERC CREATE program in photovoltaics at McMaster University for its financial support.

REFERENCES

1. M. J. Keevers, *et al.*, J. Appl. Phys. 75(8):4022-4031 (1994).
2. M. Green, *Third Generation Photovoltaics*, (Springer Series in Photonics, 2003) pp. 1-5.
3. M. Green, *Solar Cells – Operating Principles, Technology and System Applications*, (The University of New South Wales, 1998) pp. 62-83.
4. E. Huante-Ceron, PhD. Thesis, McMaster University, 2011.
5. P. Bermel, *et al.*, Opt. Express 15, 16986-17000 (2007).
6. K. Feldrapp, *et al.*, Prog. Photovolt: Res. Appl. 2003; 11:105-112.
7. J.S.C. Prentice, J. Phys. D: Appl. Phys. 33(2000) 3139-3145.

Mater. Res. Soc. Symp. Proc. Vol. 1536 © 2013 Materials Research Society
DOI: 10.1557/opl.2013.914

Impact on Thin Film Silicon Properties and Solar Cell Parameters of Texture Generated by LaserAnnealing and Chemical Etching of ZnO:Al

Rym Boukhicha[1], Erik Johnson[1], D. Daineka[1],Antoine Michel[1],
J.F. Lerat[2], Thierry Emeraud[2], and Pere Roca i Cabarrocas[1]

[1] CNRS, LPICM, Ecole Polytechnique, 91128 Palaiseau, France
[2]Excico Group NV,KempischeSteenweg 305/2, B-3500 Hasselt, Belgium

ABSTRACT

The use of a laser annealing and chemical texturing process (dubbed the LaText process) on room-temperature sputtered ZnO:Al has been shown to generate unusually high haze properties, favorable for thin film silicon solar cells.This is due to the melting of the ZnO:Al layer by the XeCl laser, and the formation of crystalline domains onthe surface, for which the grains and grain boundaries are subsequently etched at different rates. The unusual surface morphology produced through this process can strongly impact the nature of the amorphous or microcrystalline silicon material deposited thereupon. In this paper, we report on results for amorphous silicon devices, for which the surface texture is seen to slightly impact thelight absorption in the material, but more interestingly, also the light-induced degradation of the cells.For co-deposited cells, devices deposited on surfaces with the characteristic "LaText" morphologyundergo a much lesser degradation. Furthermore, the decreased degree of degradation coincides with a notable shift in the Raman scattering peak. This provides a rapid diagnostic for testing multiple textures and deposition parameters.

INTRODUCTION

Doped ZnO is used in both research and industry as the transparent conducting oxide (TCO) front contact in thin film silicon solar photovoltaic cells, as well as in optoelectronics[1]. In the superstrate configuration, the upper front contact layer should meet a number of requirements: high transparency in the visible to near-infrared solar spectrum, a high electrical conductivity, a suitable surface texture in order to enhance light scattering and absorption inside the cell, a high chemical stability, and good adhesion to silicon. To meet these goals, either sputtered ZnO:Al is used after a chemical etching step [2, 3], or low-pressure chemical vapour deposited (LP-CVD) ZnO:B [4]. In this work, we examine a third structural option, wherein the surface morphology is created by an excimer laser annealing and chemical etching technique, which we dub the LaText technique. LaText treated ZnO:Al layers have a low resistivity, high transmission in the visible range, and a high haze [5]. It has been previously shown that such a substrate treatment dramatically increases the photocurrent in hydrogenated microcrystalline silicon (µc-Si:H) solar cells [6]. In this work, we study the influence of the LaText process on the optoelectronic and light-induced degradation (LID) properties of hydrogenated amorphous silicon (a-Si:H) solar cells.

EXPERIMENT

ZnO:Al thin films were deposited on Corning Eagle glass at room temperature by radio frequency magnetron sputtering from a ceramic ZnO target mixed with Al_2O_3 (1% wt). The target diameter is about 15 cm and the process chamber was pumped to a base pressure of less than 5.10^{-7} mbar. The samples have been deposited at 0.11 Pain pure argon. After deposition, the ZnO:Al films were excimer laser annealed (ELA) using a pulsed UV exciplexXeCl laser (EXCICO LTA 15 series) with an emission wavelength of 308nm. The pulses generated by this source were longer than 150ns, and the energy per pulse was varied. Each energy dose was delivered in a single shot over a constant area of 1x1 cm^2 with a cross-beam non-uniformity of less than ±2%. In this way, up to 16 different fluences of ELA treatment have been investigated per 5x5cm² substrate. The ZnO:Al thin film samples were subsequently wet-etched in a dilute (0.5%) aqueous HClsolution for a short time of 10s, leading to a textured roughness on the surface. The chemical texturing step is similar to the one reported by the Julich group [2] to produce textured substrates from ZnO:Alsputtered at high temperatures (250–400 °C).

The PIN stacks were then prepared by Plasma-Enhanced Chemical Vapour Deposition (PECVD) at 175°C in a single-chamber commercial NEXTRAL reactor.The a-Si:H cells were composed of the following layer stack: Corning Eagle/ZnO:Al/p-a-Si:H (20nm) /a-Si:H (300nm) /n-a-Si:H (20nm)/Aluminum (evaporated). The p-a-Si:H was not optimized for good ohmic contact with the ZnO:Al, and limited the fill-factors for the devices. Furthermore, the Al back-contacts are poor IR reflectors, and limited the red response of the cells.

The Raman spectra of the layers within the cells were measured using a high-resolution (0.1 cm^{-1}) Raman spectrometer (Labram HR800 from HORIBA JobinYvon) in a confocal microscope backscattering configuration. The objective used in this study was a 100X (NA = 0.9) objective from Olympus. A tunable 532 nm wavelength laser was employed in this confocal configuration.The optoelectronic performance of the cells was studied by acquiring current-voltage (I-V) curves at AM1.5 light intensity and by external quantum efficiency (EQE) measurements over the wavelength range 400-800 nm.

DISCUSSION

Figure 1 displays the variation in PV parameters for the co-deposited a-Si:H cells as a function of the ELA fluence used on the underlying substrate. A decrease is observed in the fill factor (FF) above a threshold of about 0.5 J/cm^2 , and it further decreases from 64% to 46% as the laser fluence increases from 0.45 J/cm² up to 1 J/cm² (Fig.1-a). As previously stated, the initial low value of the FF is due to the poor contact resistance between the p-a-Si:H and the ZnO:Al. However, the decrease in FF with increasing fluence is not only due to any resistive effects in the ZnO:Al (as previously shown [6]), the p-layer, or their interface, as the series resistance read from the I-V curves does not increase until above 0.7 J/cm². The open-circuit voltage (V_{OC}) displays a decrease at the threshold of 0.45 J/cm², dropping by about 50 mV as the fluence increases to 1 J/cm². This decrease is not as important in terms of absolute value (decrease of 4%) but is still of note for co-deposited cells. On the positive side, one observes a gain in J_{SC} (> 0.5 mA/cm²) at the same threshold ELA condition of 0.5 J/cm² (Fig.1-b). It is of note that this gain is observed despite the use of aluminum back contacts (with poor red/infrared response) which minimize the light-scattering gains to be had in this low absorption region.

However, and as always for a-Si:H solar cells, devices must be light-soaked to reveal the dynamics of the devices in operation, revealing to the decreased performance of a-Si:H films through the Stabler-Wronski effect [7]. Figure 2shows the performance of a-Si:Hcells as a function of the light soaking time for a brief period of light-soaking. Each curve representsa different fluencefor the ELA treatment of the underlying substrate, and one is reminded that the cell layer stacks are co-deposited. It can be seen that for the cells on the low-fluence areas, a large degradation is observed (12% degradation in 200 hours). However, for the increased fluence areas (above the threshold of 0.45 J/cm^2 observed for the changes in the initial PV parameters), a significant reduction of the LID effect is observed. The cells on a substrate area annealed at 0.51 J/cm^2 show significantly less LID and become the highest efficiency cells after LS. Furthermore, for the highest fluence samples, no LID is observed whatsoever (although from lesser initial values).

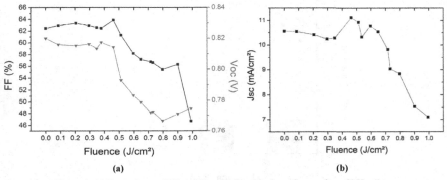

(a)   (b)

Figure 1: (a) Dependence of FF and V$_{OC}$, (b) J$_{SC}$ upon laser fluence for a-Si:H cells.

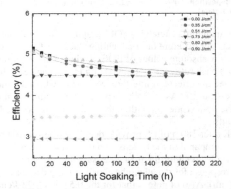

Figure 2: Variation of efficiency plotted as a function of light-soaking time for cells annealed under different laser fluences.

To make a link with the changes in material properties that may lead to such changes in cell performance and stability, the position of the TO Raman scattering peak measured for these cells is presented in Figure 3. Influenced by the underlying substrate morphology, the Raman spectra of the a-Si:H absorber layers show a shift in the position of the TO peak from 477 cm^{-1} to 475 cm^{-1} (Fig.2). It is of note that this shift occurs at the precise threshold where all optoelectronic properties shift (0.45 J/cm^2). This effect may be linked to the change in substrate morphology, occurring as the ELA treatment melts a thin surface layer. The surface changes from a granular continuous surface to a mosaic-like one with increasing laser fluence. Atomic Force Microscopy (AFM) images of the a-Si:H surface after deposition are included as insets in Figure 3, and it is clear that that the surfaces of the devices retain a memory of the underlying substrate.

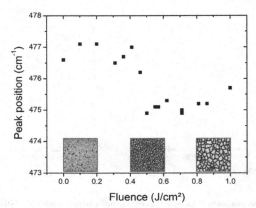

Figure 3: a-Si:H Raman shift position depending on the laser fluence. Insets: AFM images of surfaces after a-Si:H deposition, for low fluence (0.3 J/cm²), in the drop region (0.5 J/cm²) and at high fluence (0.7 J/cm²).

Figure 4 shows the external quantum efficiency (EQE) graphs before and after LS for cells deposited on substrates after four different laser annealing fluences: 0 J/cm², 0.35 J/cm², 0.54 J/cm², and 0.72 J/cm². The low absolute values of EQE are due to the thick un-optimized p-a-Si:H layer and the aluminum back reflector, but interesting trends can nevertheless be noted. For figs 4(a) and 4(b), one can see the effects of interference fringes, which are then suppressed by increased texture in figs 4(c) and (d). As well the response at 600 nm is seen to increase with increasing fluence due to the light scattering, although the discrete filters used for EQE make this observation difficult to quantify.

The changes in EQE after 180 h of LS provide details on the substrate-dependent degradation of these cells. Compared to the 0 J/cm² and 0.35 J/cm² samples, a more stable EQE over most of the spectrum is observed for the 0.54 J/cm²sample. Furthermore, for all substrates, it is seen that although the response for wavelengths above 550 nm decreases, it increases at shorter wavelengths (<450 nm). Also of note is that for the 0.54 and 0.72 J/cm² devices, although the EQE (and therefore the J_{SC}) decreases, the efficiency of the devices is absolutely stable, as the FF and V_{OC} actually increase with light-soaking (not shown).

DISCUSSION

It is well-known that the texture of the substrate changes the optoelectronic properties of a-Si:H solar cells [8].However, the impact on the decrease in efficiency (caused by the light-induced creation of dangling bonds known as Staebler-Wronski effect (SWE) [7]) has been rarely discussed. It is clear from the results presented in this work that the underlying texture has a strong effect on both the initial cell efficiency, as well as the cell stability. The increased photocurrent due to greater optical haze is an obvious effect, and the decreased V_{OC} due to increased surface area (and thus increased J_0) is also an effect that has been noted in the literature for very rough substrates [10], along with decreases in FF. We note again that the decrease in FF with increasing fluence in these cells is caused by increased recombination, not a greater R_S.

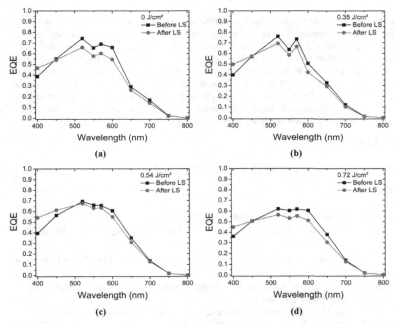

Figure 2: External quantum efficiency of cells on substrates (a) without laser annealing, and having received (b) 0.35 J/cm², (c) 0.54 J/cm², and (d) 0.72 J/cm². EQEs measured before and after light soaking.

However, for the light-soaking behavior, the reduction of the obvious effects of LID for higher fluences is remarkable. It must be noted that it is not due to a reduced exposure to light for some a-Si:H layers, as the EQE curves are of the same order of magnitude (and in fact J_{SC} increases for the highest stable efficiency cell). Furthermore, it is indicated that the underlying substrate texture is fundamentally modifying the nature of the material, as put into evidence by the change in the Raman peak location coinciding with the threshold for the formation of domains and cracks. Given this information, two explanations are imaginable: one is the growth of the silicon in the cracks and valleys produced by ELA treatment and successive etching of the ZnO – a geometry that will restrict the volume expansion of a-Si:H (a necessary component of LID [9]) and therefore reduce the creation of the defects that cause the SWE. A second possibility is that the growth of the a-Si:H on the extremely textured surface results in built-in defects that limit the FF more than the additional defects caused by the SWE, but that for the cells presented, the limitation of the built-in defects is compensated by the higher current due to light-scattering, resulting in a net gain for these cells. Further experiments are required to quantify if the LaText technique can lead to a net stabilized gain in a-Si:H cell efficiency.

CONCLUSIONS

In this work, we studied the beneficial effect of LaText-treatedZnO:Alon the performance and stability of a-Si:H photovoltaic cells. The best, stabilized cells were obtained on ZnO:Al laser annealed at around 0.5 J/cm² due to a combination of improved light-trapping and less light-induced degradation. For cells deposited on material annealed at the highest fluences, no decrease in cell efficiency was observedduring light-soaking. Raman scattering shows that the cell stability coincides with a shift in the peak location to lower wavelengths. Such results are very promising given the direct improvements that can be done to improve the performance of the cell, such as using intrinsically more stable material, using a p-layer with better contact to the ZnO:Al, and using a better back-reflector such as ZnO:Al/Ag.

ACKNOWLEDGMENTS

This work has been supported by the FUI Project "LaText".

REFERENCES

1. T. Minami, **Semicond. Sci. Technol. 20**, (2005) S35.
2. C. Agashe, O. Kluth, J. Hupkes, U. Zastrow, and B. Rech, **J. Appl. Phys. 95** (2004) 1911.
3. Berginski, M.; Hupkes, J.; Reetz, W.; Rech, B. and Wuttig, M. **Thin Solid Films 516** (2008) 5836.
4. S. Fay, L. Feitknecht, R. Schluchter, U. Kroll, E.Vallat-Sauvain, and A. Shah, **Sol. Energy Mater. Sol. Cells 90** (2006) 2960.
5. E. V. Johnson, C. Charpentier, T. Emeraud, J.F. Lerat, C. Boniface, K. Huet, P.Prod'homme, and P. Roca iCabarrocas, **Amorphous and Polycrystalline Thin-Film Silicon Science and Technology - 2011 (Mater. Res. Soc. Symp. Proc.)** A13.5.
6. E: V. Johnson , P. Prod'homme , C. Boniface , K. Huet , T. Emeraud , and P. Roca I Cabarrocas, **Sol. Energy Mater. Sol. Cells 95**, (2011) 2823.
7. Staebler, D. L. and Wronski, C. R. **Appl. Phys. Lett. 31** (1977) 292.
8. Hegedus, S.; Buchanan, W.; Liu, X.; Gordon, R., **Photovoltaic Specialists Conference, 1996., Conference Record of the Twenty Fifth IEEE** (1996) 1129.
9. P. Tzanetakis, **Sol. Energy Mater. Sol. Cells. 78**, (2003) 369.
10. M. Boccard, P. Cuony, M. Despeisse, D. Dominé, A. Feltrin, N. Wyrsch,and C. Ballif, **Sol. Energy Mater. Sol. Cells 95**, 195–198 (2011).

Mater. Res. Soc. Symp. Proc. Vol. 1536 © 2013 Materials Research Society
DOI: 10.1557/opl.2013.743

Novel intermediate reflector layer for optical and morphological tuning in the Micromorph thin film tandem cell

J.-B. Orhan[1], E. Vallat-Sauvain[1], M. Marmello[1], U. Kroll[1], J. Meier[1], E. Laux[2], D. Grange[2], S. Farine Brunner[2], H. Keppner[2]

[1]TEL Solar-Lab S.A., rue du Puits-Godet, 12a, 2000 Neuchâtel, Switzerland
[2] Haute Ecole Arc Ingénierie, Eplatures-Grise 17, 2300 La Chaux-de-Fonds, Switzerland

ABSTRACT

PECVD growth of the microcrystalline silicon junction on a highly textured amorphous top cell often leads to defective absorber layers and finally to low quality bottom cell. This paper reports on the current status of using an innovative smoothening/reflective layer (SRL) as alternative intermediate reflector between top and bottom cell of a Micromorph tandem device deposited on as-grown highly textured LPCVD ZnO layer. Manufacturing of the SRL layer is realized by "liquid phase" deposition technologies. Optical and electrical properties, smoothening effect and photoelectrical results of Micromorph tandem devices are discussed. The implementation of our novel SRL results in the growth of a crack-free bottom cell and to an efficient current transfer from the bottom to the top cell.

INTRODUCTION

TEL Solar is active in the field of production equipment for silicon based thin film PV modules, providing turn-key thin film production lines for a-Si and, a-Si/μc-Si tandem technologies [1-2]. To achieve the goal of high efficiency, TEL Solar uses a proprietary transparent conductive oxide (TCO) for the coating of the glass which is based on zinc oxide (ZnO) deposited by low pressure chemical vapor deposition (LPCVD). Its advantage lies in the fact that it is rough as grown and, therefore, it acts as an excellent light diffuser. As a result, the absorption of the light in the silicon layers increases, generating high current density of the solar cell. If even thicker and rougher TCO layers could be used, the current density would increased further [3]. Such layers would however lead to devices with low V_{oc} and FF. The formation in the bottom microcrystalline cell of low-density defects [4], so-called cracks, probably acting as shunts and recombination centers, is believed to be the origin of the photoelectrical properties degradation. We report on our first results in the development of a novel interlayer to be inserted between the top and bottom cells of a Micromorph tandem device. This layer presents two main functionalities: on one hand it levels the surface by lowering its roughness; on the other hand it acts, thanks to its low refractive index, as an optical reflector. Levelling allows the growth of a better quality μc-Si bottom cell. The reflection of light improves the current in the top amorphous cell. Therefore, this interlayer is expected to allow the use of highly texture LPCVD ZnO as front contact, and hence to reach high currents and high efficiencies. The manufacturing of this layer is proposed using inventive "liquid phase" deposition technologies. Furthermore, "liquid phase" deposition equipments and consumables are, economically speaking, advantageous compared to plasma enhanced chemical vapor deposition (PECVD) ones. Additionally, a smoothening intermediate reflective layer would allow to design completely new type of tandem devices as a-Si top cells are more resilient to rough surfaces than μc-Si:H bottom cells. SRL optical and electrical properties, smoothening effect and photoelectrical results of Micromorph tandem devices are discussed in this paper.

EXPERIMENTAL DETAILS

SRL, based on commercially available products, were deposited using standard spin coating procedures followed by the required drying and annealing steps. All involved process temperatures are below 200°C. SRLs thicknesses reported in this paper refer to thicknesses of SRLs deposited on polished crystalline silicon wafers.

The tandem cells presented in this work have been deposited on an in-house as-grown rough LPCVD ZnO front TCO displaying a high haze factor of ~60% at 600 nm. Amorphous top cells, microcrystalline bottom cells and silicon oxide based intermediate reflectors (SOIR, alternatively called n-SiOx) have all been fabricated in the TEL Solar KAI-M PECVD reactor at an excitation frequency of 40.68 MHz. Back contact was realized by LPCVD ZnO. Test cells, without additional back reflector, were patterned by laser scribing to squares of 0.25 cm^2. The current–voltage characteristics of the cells were measured under a dual lamp WACOM solar simulator in standard test conditions (25 °C, AM1.5 spectrum, and 1000 W/m^2). External quantum efficiency (EQE) measurements were realized using an in-house setup.

Optical total transmission (T_T) and reflection (R_T) have been measured by a Perkin Elmer lambda 950 spectrometer equipped with an integration sphere. Ellipsometric measurements have been done with a spectroscopic ellipsometer Jobin Yvon – Horiba HR460. Planar conductivity was determined from coplanar measurement of layers deposited on flat glass. Scanning electron microscopy (SEM) was realized on polished cross section of devices with a Philips XL30 ESEM-FEG. Atomic force microscopy (AFM) of 10 x 10 μm surfaces were taken with 512 x 512 data points with a Veeco Bruker Icon. The average half-opening angle is calculated from AFM data according to Nasuno [5].

RESULTS AND DISCUSSION

SRL optical and electrical characterization

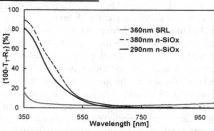

Figure 1: Absorptance, defined by (100-T_T-R_T), of PECVD 290 and 380 nm n-SiOx, and of 360 nm SRL measured on glass against air.

Typical absorptance curves, defined by $(100-T_T-R_T)$, are shown in Figure 1. For $\lambda < \sim 650$ nm, the SRL presents a significantly lower absorption than a standard n-SiOx. For longer wavelength, absorption is slightly higher for the SRL. Ellipsometric characterization of such layers gives for $\lambda = 600$ nm refractive indexes of: $n_{n-SiOx} \approx 1.8$ and $n_{SRL} \approx 1.6$. The planar conductivity measurement of our SRL results in $\sim 10^{-2}$ S.cm^{-1}. As our material is believed to be isotropic, we assumed our transverse conductivity to be comparable to the planar one. With such optical and electrical properties, our SRL fulfills the basic requirements for utilization as intermediate reflective layer.

Morphological characterization

To characterize the smoothening ability of our SRL in real cells, the test substrate used was "glass + LPCVD ZnO + a-Si top cell". Figure 2 presents the 3-D imaging of AFM measurements of a) a reference sample and of b) a sample with 330 nm of our SRL. Results of the statistical analysis of the AFM data are reported in Table I.

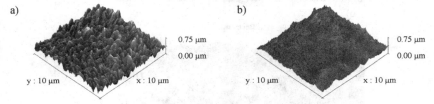

Figure 2: 3-D imaging of AFM measurement of the surface of a) glass + LPCVD ZnO + a-Si top cell, b) glass + LPCVD ZnO + a-Si top cell + 330 nm SRL

Table I: Statistical analysis of AFM measurements corresponding to Figure 2.

	R$_{RMS}$ [nm]	Peak to Valley height [nm]	Half-opening angle [°]
ZnO + Top cell	100	708	45
ZnO + Top cell + 330 nm SRL	21	235	83

The significant lowering of the RMS roughness and of the peak to valley height demonstrates an efficient smoothening of the underlying top cell surface. Moreover, by applying the SRL concept, the increase of the half-opening angle indicates an opening of the valley.

Tandem device

Figure 3 presents SEM pictures of polished cross-sections of different Micromorph devices. A simple, i.e. without SOIR, tandem cell is presented in Figure 3a). Many "cracks" can be seen in the bulk of the microcrystalline cell. Those "cracks" [4-6] originates especially from sharp and deep valleys present on the "ZnO + a-Si top cell" surface. The presence of such defects leads to a poor performance of the bottom μc-Si cell and, hence, to a poor overall performance of the Micromorph tandem device.

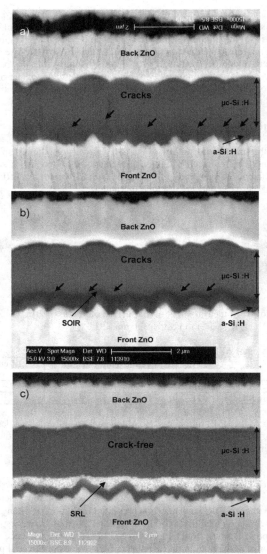

Figure 3: Polished cross-section SEM pictures of a) Micromorph device, b) Micromorph device with 300 nm SOIR, c) Micromorph device with 390 nm SRL. Cracks indicated with an arrow.

Same observations can be made in a Micromorph device with an additional 300 nm thick SOIR as presented in Figure 3b). Indeed, a conformal growth of the n-SiOx layer is observed. This leads to the conservation, with only a very shallow smoothening, of the morphology "ZnO + a-Si top cell" surface. A Micromorph device with a 390 nm SRL is shown in Figure 3c). A significant flattening of the "ZnO + a-Si top cell" surface is observed. This is due to the liquid deposition method that preferentially fills the valley of the underlying rough surface. This smoothening leads to the growth of a bottom microcrystalline device without "cracks". A noticeable flattening of the microcrystalline cell / back ZnO interface is also observed when our SRL is introduced in the Micromorph device.

The electrical characterization of our first device with SRL is presented in Figure 4. For comparison purpose, additional I-V and EQE response of Micromorph devices without and with 300 nm SOIR are presented. No particular adaptation of the PECVD processes was used for the integration of the SRL.

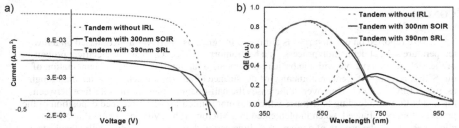

Figure 4: a) I-V and b) EQE properties of Micromorph devices, without/with SOIR and with SRL

On Figure 4a), the device with SRL presents a significantly too high open circuit resistance (R_{oc}). We believe that it comes from the not optimized interfaces between the PECVD layers and the SRL. Regarding the device with SOIR, a R_{oc} increase is also observed. It can be explained by the low transverse conductivity ~10^{-5} S.cm^{-1} of such thick n-SiOx layers [6]. The low short circuit resistance observed for this sample is not understood. EQE of the devices, shown in Figure 4b), indicate an efficient current transfer from the bottom to the top cell. Compared to tandem device with 300 nm SOIR, together with a higher top cell response between $600 < \lambda < 750$ nm, a higher response of the bottom cell is observed between $500 < \lambda < 650$ nm. We believe that the higher top response originates from the lower refractive index of the SRL. The higher response of the bottom cell for $500 < \lambda < 650$ nm is believed to be linked with the higher transparency of the SRL in this range as illustrated in Figure 1. In the bottom cell, for $\lambda > 650$ nm, the lower response can be attributed to a combination of two different phenomenon: first to the reduction of light scattering due to the flattened interface illustrated in Figure 3c) and second to the higher SRL absorption in this wavelength range as illustrated in Figure 1. The relative influence of each phenomenon is not yet quantified. Spectral responses data are also summarized in Table II by the corresponding calculated current densities. At identical device total current of 19.5 mA/cm^2, the 390 nm SRL allows a more efficient current transfer towards the top cell of + 2.7 mA/cm^2 than a device with 300 nm SOIR having an increase of only + 2.4 mA/cm^2.

Table II: Current calculated from EQE of Micromorph devices, without/with SOIR and with SRL

J_{sc} from EQE [mA/cm^2]	Top	Bottom	Top+Bottom
Tandem without IRL	11.7	11.0	22.8
Tandem with 300nm SOIR	14.1	5.5	19.5
Tandem with 390nm SRL	14.4	5.1	19.5

In summary, when introduced in a Micromorph device as IRL, a 390 nm thick SRL leads to the growth of a μc-Si bottom cell without cracks. It also allows an efficient current transfer from the bottom to the top cell. Nevertheless, the integration of this new layer is not yet optimized as can be seen in the I-V curves of Figure 4a).

CONCLUSIONS

We reported on our first results aiming to integrate as interlayer a SRL, based on a low temperature sol-gel deposition process, in a Micromorph device. Optical and electrical properties of the SRL are compatible with the basic requirements of an interlayer. The implementation of our SRL allows an efficient smoothening of the surface of an a-Si top cell deposited on a rough as grown LPCVD ZnO. Moreover, it increases the half-opening angle of the interface between top and bottom cell leading to textures which are more suited for crack-reduced deposition of the μc-Si bottom cell. Indeed, when implemented in a Micromorph device, a 390 nm thick SRL allows, as aimed, the growth of a microcrystalline cell without cracks. Additionally, thanks to a refractive index of 1.6, it leads to the desired efficient current transfer from the bottom to the top cell. Thus, a good optical integration is demonstrated. Nevertheless, the electrical integration is not yet fully optimized and will be addressed in the next months. On-going work is focused on the consolidation of this proof of concept, as well as on the optimized electrical integration of the SRL in Micromorph devices.

ACKNOWLEDGMENTS

Research supported by the Swiss Commission for Technology and Innovation (CTI) under project # 12614.1 PFIW-IW. The authors would like to thank M. Leboeuf and M. Dadras from CSEM for support with AFM and SEM imaging.

REFERENCES

1. S. Benagli et al., Proc. of 24th EU-PVSEC (2009)
2. J. Bailat et al., Proc. of 25th EU-PVSEC (2011)
3. J. Steinhauser, L. Feitknecht et al., Proc. of the 20th EU-PVSEC, 1608 (2005).
4. D. Dominé, PhD Thesis, University of Neuchatel (2009)
5. Y. Nasuno et al. Jpn. J. Appl. Phys. 40, L303 (2001)
6. P. Cuony, PhD thesis, EPFL (2011)

Mater. Res. Soc. Symp. Proc. Vol. 1536 © 2013 Materials Research Society
DOI: 10.1557/opl.2013.738

Development of Nanocrystalline Silicon Based Multi-junction Solar Cell Technology for High Volume Manufacturing

Xixiang Xu[1], Jinyan Zhang[1], Anhong Hu[1], Cao Yu[2], Minghao Qu[1,3], Changtao Peng[1], Xiaoning Ru[1], Jianqiang Wang[2], Furong Lin[2], Hongqing Shan[1], Yuanmin Li[1], and Hui Yan[3]

[1]Apollo Precision Equipment Limited Company, R&D Center, Shuangliu, Sichuan, China
[2]Hanergy Holding Group Limited, Beijing, China
[3]Colleage of Materials Sci., Beijing Univ. of Technology, Beijing, China

ABSTRACT

We conduct a comparative study mainly on two types of nc-Si based solar cell structures, a-Si/a-SiGe/nc-Si triple-junction and a-Si/nc-Si double-junction. We have attained comparable initial efficiency for the both solar cell structures, 10.8~11.8% initial total area efficiency (85 - 95W over an area of 0.79 m^2). For better compatibility to our installed manufacturing equipment, we deposit a-Si and a-SiGe component cells with the existing deposition machines. Only nc-Si bottom component cells are prepared in separate deposition machines tailored for nc-Si process. Material properties of nc-Si and TCO films are also studied by Raman spectra, SEM, and AFM.

INTRODUCTION

Thin film silicon, amorphous silicon (a-Si) and nanocrystaline silicon (nc-Si), has evolved into an important technology for photovoltaic industry in the last decade. Over 2 GW capacities of amorphous silicon and silicon-germanium (a-SiGe) multijunction solar cell manufacturing lines using Apollo/Hanergy's turn-key technology have been installed in Hanergy Group and GS-Solar factories. However, compared with other thin film PV technologies, such as CdTe and CIGS with 11-14% product efficiency, most of thin film Si product's efficiency is still significantly lower, in the range of 8-10%. Therefore, evolution of thin film silicon PV technology shows a trend moving from a-Si and a-SiGe multi-junction structures, showing light-induced degradation in range of 10-20%, to a more stable nc-Si based ones with typical ~10% or less light-induced degradation [1-4]. Hybridizing a-Si and a-SiGe with nc-Si, Yan et al. reported 16.3% initial efficiency for a small area a-Si/a-SiGe/nc-Si triple-junction cell (0.25 cm^2) [5].

In order to increase cell conversion efficiency of our thin film Si PV products, we have focused our effort on development of compatible nc-Si technology. We have conducted our experiments mainly on two types of nc-Si based solar cell structures, a-Si/a-SiGe/nc-Si triple-junction and a-Si/nc-Si double-junction device, with schematic cell structures illustrated in Fig. 1. Currently we are attaining slightly higher initial efficiency for the a-Si/nc-Si double-junction than a-Si/a-SiGe/nc-Si triple-junction structure, 93W vs. 85 W, corresponding to 11.8% and 10.8% initial total area (0.79 m^2) efficiency.

Other experimental results, including study of volume fraction of crystalline (Raman spectroscopy) along nc-Si growth, individual component cell optimization and current match assisted by QE (quantum efficiency), and development of superior tunnel-junction and contact layers, will also be discussed.

EXPERIMENTAL

All samples reported in this paper are prepared in R&D center in Shuangliu, Sichuan Province, China. For better compatibility to the installed manufacturing equipment, we fabricate a-Si and a-SiGe component cells with the existing deposition machines. Only nc-Si bottom component cells are prepared in separate deposition machines tailored for nc-Si process. In order for smooth technology transfer, we run all thin film Si deposition on product size glass substrate (1.245m x 0.635m). Amorphous Si and SiGe component cells are deposited using conventional RF (13.56 MHz) PECVD machines in a batch mode. Each run can load 72-piece of 0.79 m^2 glass. nc-Si bottom cells are fabricated in a VHF (40 MHz) excited PECVD depositor.

IV characteristics are mainly measured with Pasan, Spire, and DLSK solar simulators in full size (with total area of 0.79 m^2, and active area close to 0.73 m^2). Other measurements, such as quantum efficiency (QE), thickness, and crystalline volume fraction are also conducted to evaluate component cell performance and spatial uniformity.

Two types of experiments are also conducted to evaluate light stability, indoors and outdoors. For indoor light soaking, full-size encapsulated samples are exposed to simulated light of one sun intensity in open-circuit mode under 50 °C. For the outdoor light exposure, the samples are soaked in our outdoor light soaking site in Panzhihua, Sichuan, with good annual solar insolation and better spectrum close to AM1.5.

RESULTS AND DISCUSSIONS

In this work, we will mainly report experimental progress on a) optimization of nc-Si process with deposition rate in range of 3 – 6 Å/sec, b) tunnel-junction improvement between a-Si or a-SiGe and nc-Si component cells, c) comparison of FTO coated glass and home developed B-doped ZnO (BZO), and d) IV characteristics of a-Si/nc-Si double-junction in comparison of a-Si/a-SiGe/nc-Si triple-junction devices.

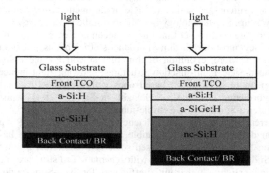

Figure 1. Schematic of two types of nc-Si based multi-junction solar cells studied in this work: a-Si/nc-Si double- and a-Si/a-SiGe/nc-Si triple-junction.

nc-Si:H deposited at different rate

Deposition rate of intrinsic nc-Si layer is usually the bottleneck of manufacturing throughput since it is the thickest layer in the nc-Si based multi-junction solar cell. We first developed nc-Si deposition process at ~3A/sec, and obtained initial total area efficiency close to 12%. With optimizing process parameters, such as reactive gas flow, pressure, and VHF power, the cell efficiency almost remains constant up to ~6A/sec. Initial IV characteristics of three representative samples are summarized in Table I with nc-Si deposition rate at 3.4A/sec, 4.4A/sec, and 5.9A/sec, respectively.

Table I. Initial IV results of large-area (total area 0.79 m^2) a-Si/nc-Si double-junction modules made at different deposition rate.

Run #	Dep. rate	Pmax (W)	Isc (A)	Voc (V)	FF	Efficiency (%)
580	3.4A/s	93.0	2.37	54.0	0.727	11.8
593	4.4A/s	91.5	2.37	53.3	0.724	11.6
603	5.9A/s	91.2	2.36	53.5	0.722	11.6

Raman spectra of nc-Si films

Film properties such as crystalline volume fraction and its evolution along film growth of nc-Si layer can be effectively monitored by Raman spectra. The 633 nm Raman spectra of a set of nc-Si thin films interrupted at different thicknesses are illustrated in Figure 2. The crystalline volume fraction is in the range of 60 -67% when the nc-Si grows from 393 nm to 1984 nm. The lower crystalline volume fraction for the 2523 nm nc-Si is caused by an intentional grading.

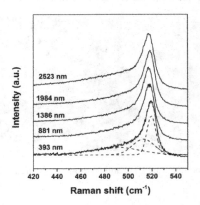

Fig.2. Raman spectra of 5 nc-Si films with different thicknesses, which can probe the nc-Si growth and vertical evolution.

Tunnel-junction optimization

Effect of tunnel-junction, between nc-Si and a-Si cells, on a-Si/nc-Si double-junction cell performance is summarized in Table II. With a superior tunnel-junction (Type I), FF is significantly improved due to reduced series resistance and better component cell current match. Compared with Type III, Type I tunnel-junction improves FF by ~7%.

Table II. I-V characteristics of a-Si/nc-Si double-junction solar panels with different tunneling-junction structures.

Expt #	Tunnel-junction	Pmax (W)	Isc (A)	Voc (V)	FF	Efficiency (%)
1	Type I	89.93	2.265	53.56	0.741	11.4
2	Type II	88.80	2.273	54.09	0.722	11.2
3	Type III	85.76	2.366	52.19	0.695	10.9

FTO vs. BZO contact layers

TCO films used for either or both front-side and back-side electrodes of Si thin film solar cells can effectively enhance photon absorption, resulting in an increase in the cell current density. We did a comparative study on the surface morphology and texture features of different types of TCO by SEM and AFM. As shown in Figure 3 (a) and (b), the home-deposited B-doped ZnO (BZO) shown in (b) demonstrates a more advantageous features than commercial FTO shown in (a), as the front side electrodes to enhance light absorption.

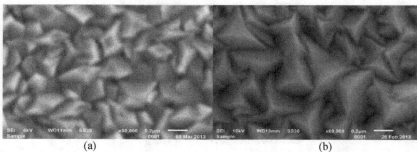

(a)   (b)

Figure 3. SEM images of (a) commercial FTO and (b) home-deposited BZO films.

Figure 4 shows QE of three a-Si/nc-Si double-junction cells with different front-side contact, two with FTO coated glasses and one with home-developed BZO. The sample using FTO2 shows higher QE than FTO1 due to higher transmission and better light trapping. The cell deposited on BZO coated glass demonstrates the highest Jsc mainly because of its improved QE in long wavelength range beyond ~650 nm, resulting in higher current in the nc-Si bottom cell. The higher QE in the range of shorter wavelength (<650 nm) for the cells deposited on BZO is not clear yet, which may result from a combination of differences in *p*-layer quality and different glass makers.

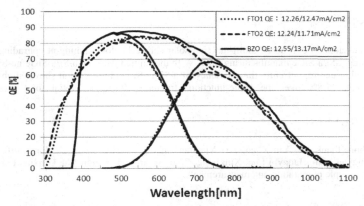

Fig. 4.   Quantum efficiency (QE) of a-Si/nc-Si double-junction cells deposited on different front-side TCO's.

Double- vs. triple-junction cell structure

As illustrated in Table I, we have attained 11.8% initial total area efficiency on the a-Si/nc-Si double-junction sample. One natural approach to achieve higher efficiency is to stack another component cell to form a triple-junction cell structure [3]. We conducted some experiments with a-Si/a-SiGe/nc-Si triple-junction structures. Based on our limited experiments, we obtained 10.7% initial total area efficiency for a-Si/a-SiGe/nc-Si triple-junction, which is lower than 11.8% for a-Si/nc-Si double-junction. The limiting component cell is a-SiGe middle cell.

We also conducted light soaking on the a-Si/nc-Si double-junction and a-Si/a-SiGe/nc-Si triple-junction panels, both in-door and outdoors (at Panzhihua, Sichuan). Because not optimized current match between component cells and possible large degradation in a-SiGe middle cells, the triple-junction samples typically show 2-3% more light-induced degradation than the double-junction samples, ~15% vs. ~12%. It is conceivable that with further optimization, the gap in the light-induced degradation between the double- and triple-junction structures can be narrowed.

Table III. Initial I-V characteristics of large-area a-Si/nc-Si double- and a-Si/a-SiGe/nc-Si triple-junction panels as well as their corresponding QE's of component cells. FTO (labeled as FTO1 in Fig. 4) is used for both double- and triple-junction cells.

Sample #	Cell type	Pmax (W)	Isc (A)	Voc (V)	FF	Eff. (%)	QEtop (mA/cm^2)	QEmid (mA/cm^2)	QEbot (mA/cm^2)	QEtotal (mA/cm^2)
1	Double	93.03	2.37	54.0	0.73	11.8	12.26		12.47	24.71
2	Triple	84.77	1.39	81.0	0.75	10.7	7.55	7.32	7.94	22.81

CONCLUSIONS

We have developed nc-Si based multi-junction solar cell technology aiming at upgrading our current existing a-Si/a-SiGe manufacturing lines. With the limited work on a-Si/a-SiGe/nc-Si tirple-junction cell structure, currently the efficiency of a-Si/nc-Si double-junction shows slight superior efficiency, 11.8% over the triple-junction structure, 10.8%, of initial total area efficiency.

ACKNOWLEDGMENTS

The authors thank Dr. Yin Zhao, Dr. Xiaodan Zhang, and Dr. Song He for fruitful discussion, Chongyan Lian, Bangyin Lan, Xinghong Zhou, Shihu Lan, Zhijun Wu, Guixian Lei, and Yun Han for sample preparation and measurements.

REFERENCES

1. J.Meier, R. Fl"uckiger, H. Keppner, and A. Shah, Appl. Phys. Lett. , 6 , 860(1994).
2. J. Yang, A. Banerjee, S. Guha, Appl. Phys. Lett., 70 , 2975(1997).
3. X. Xu, T. Su, S. Ehlert, G. Pietka, D. Bobela, D. Beglau, J. Zhang, Y. Li, G. DeMaggio, C. Worrel, K. Lord, G. Yue, B. Yan, K. Beernink, F. Liu, A. Banerjee, J. Yang, and S. Guha, "Large area nanocrystalline silicon based multi-junction solar cells with superior light soaking stability", in *35th IEEE PVSC*, Hawaii, July, 2010.
4. X. Xu, J. Zhang, D. Beglau, S. Ehlert, Y. Li, G. Pietka, G. Yue, B. Yan, C. Worrel, A. Banerjee, J. Yang, and S. Guha, "High efficiency large-area a-SiGe:H and nc-Si:H based mulit-junction solar cells: a comparative stdy," in *25th European Photovoltaic Solar Energy Conference and Exhibition,* 2010, p.2783.
5. B. Yan, G. Yue , L. Sevic, J. Yang, and S. Guha, Appl. Phys. Lett. 99, 113512 (2011).

Mater. Res. Soc. Symp. Proc. Vol. 1536 © 2013 Materials Research Society
DOI: 10.1557/opl.2013.744

SnO$_2$:F with Very High Haze Value and Transmittance in Near Infrared Wavelength for Use as Front Transparent Conductive Oxide Films in Thin-Film Silicon Solar Cells

Masanobu Isshiki[1,2], Yasuko Ishikawa[1], Toru Ikeda[1], Takuji Oyama[1], Hidefumi Odaka[1], Porponth Sichanugrist[2] and Makoto Konagai[2,3]

[1] Research Center, Asahi Glass Co., Ltd., 1150 Hazawa-cho, Kanagawa-ku, Yokohama-shi, Kanagawa 221-8755, Japan
[2] Department of Physical Electronics, Tokyo Institute of Technology, 2-12-1, Oookayama, Meguro-ku, Tokyo 152-8552, Japan
[3] Photovoltaic Research Center (PVREC), Tokyo Institute of Technology, 2-12-1, Oookayama, Meguro-ku, Tokyo 152-8552, Japan

ABSTRACT

Low sheet resistance (high mobility) with high transmittance in all wavelength is required for front TCO. High haze value is also required for effective light trapping. For this purpose, we have combined F-doped SnO$_2$ (FTO) with high mobility deposited by LPCVD and reactive ion etching (RIE) processed glass substrate. However, two problems have been found. (1) The mobility of FTO on RIE substrate dropped from that on flat glass (75 to 36 cm^2/Vs). To avoid this drop, thicker film is needed. (2) To keep high transmittance with thicker film, lower carrier concentration is needed. But the mobility dropped with lower carrier concentration. In order to solve these constrains, we have adopted a stacked structure using thick non-doped layer of 2700 nm and thin F-doped layer of 500 nm. With this novel approach, we have successfully achieved the high mobility (80 cm^2/Vs), low carrier concentration (2.2x10^{19} /cm^3) and high haze value (77% at wavelength of 1000 nm) at the same time. This new developed high-haze SnO$_2$ is a new promising TCO for thin-film Si solar cells.

INTRODUCTION

For front transparent conductive oxide (TCO) films used in thin-film silicon solar cells, high transmittance in all wavelength, low sheet resistance and high haze value are required. To achieve high transmittance and low sheet resistance, the carrier concentration and mobility of TCO need to be low and high, respectively. Fig. 1 shows simulation results based on Drude model [1]. From Fig. 1(a), the absorption is almost the same in various carrier concentration if the mobility and the product of carrier concentration and film thickness (n*d) are fixed. From Fig. 1(b), the absorption becomes higher rapidly as mobility decreases. So, higher mobility is very important to achieve lower absorption especially in the longer wavelength.

TCOs with high mobility of around 80 cm^2/Vs have been reported by several groups including ours [2-4]. But all of them have flat surfaces and low haze values. Recently, W-textured ZnO films fabricated by metal–organic chemical vapor deposition (MOCVD) on reactive ion etching (RIE) processed glass substrates are reported to have very high haze value [5]. However, the mobility of MOCVD B-doped ZnO is not high enough. Therefore, we have developed F-doped SnO$_2$ (FTO) with higher mobility and higher haze value using RIE-etched substrate.

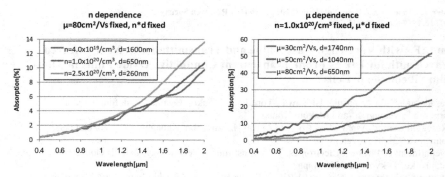

Fig. 1: (a) n dependence. μ is fixed. n*d is fixed constant. (b) μ dependence. n is fixed. So, μ*d is fixed.

EXPERIMENT

In all experiments, 0.7-mm-thick alkaline-free glass substrates (Corning 7059) were used. Before FTO-film deposition, these substrates were etched in SAMCO RIE-10NR by a standard RIE process using carbon tetrafluoride (CF_4) as the etching gas. The glass etching time was varied from 0 to 90 min. Power, CF_4 flow rate and pressure during RIE are set as 200 W, 15 sccm and 7 Pa, respectively. FTO was deposited on both RIE and flat substrates using low pressure chemical vapor deposition (LPCVD). Detail of the LPCVD deposition machine is described in Ref [2]. $SnCl_4$ and H_2O were vaporized using the bubbling method. N_2 was used as the carrier gas. HF, which was used to dope fluorine into SnO_2, was vaporized directly from liquid HF. These three gases were mixed downstream of the showerhead to avoid the pipe clogging. The calculated flow rates (excluding that of the carrier N_2 gas) of $SnCl_4$, H_2O, and HF were set at 2, 400 and 0~8 sccm, respectively. The deposition period was varied from 20 to 90 min. In all the experiments, the chamber pressure was kept at 100 Pa during deposition. During deposition, the temperature of the glass substrate was maintained at 360 or 380°C. All samples were annealed at 400°C in 100% N_2 (1 atm) for 10 min before evaluation. We have tried two film structures. One is consisted entirely of F-doped layer while the other is stacked structure of non-doped layer (substrate side) and F-doped layer (top side). Non-doped and F-doped layers are deposited continuously by only changing HF flow from 0 to 8sccm.

Film thickness on flat substrates were measured using a DEKTAK M6, which is a common use apparatus. Film thickness on RIE substrates were not measured. However, since the same conditions were used for FTO deposited both on RIE and flat substrates, their film thickness on RIE and flat substrates were supposed to be the same. In fact, we confirmed that the thickness on RIE and flat substrates with same deposition conditions were same by SEM cross-sectional observation. The electrical resistivity ρ, the carrier concentration n and the mobility μ of the films were obtained by Hall measurements in the van der Pauw configuration using BioRad HL5500.

Transmittance (T), reflectance (R), absorption(A) and haze ratio were measured by PerkinElmer Lambda950 spectrometer. It is difficult to measure T, R, and A of textured FTO films accurately because the light will be scattered by the textured surface. T, R, and A of FTO films on flat substrates can be measured accurately by adopting IM technique[6], which suppresses the light scattering effect at FTO films /air interface. However, it is impossible to measure T, R, and A of FTO films on RIE substrates accurately even if IM technique is adopted, because light is scattered at both FTO films / air and glass / FTO films interfaces. For comparison reason, IM technique has also been applied to FTO films fabricated on flat substrates. As explained in the introduction, free carrier absorption can be related to the product of carrier concentration and film thickness (n * d). We have confirmed that n * d and absorption have a linear correlation based on the measured data of FTO films on flat substrates. Here, we assumed that we can also estimate the absorption of FTO films on RIE substrates from n * d obtained by Hall measurements.

Cross-sectional SEM images were observed with Hitachi SU-70. Samples were cleaved and coated with osmium and platinum before SEM observation. The thicknesses of osmium and platinum were approximately 3 and 1 nm, respectively.

DISCUSSION

Properties on RIE substrates

Fig. 2(a) shows the mobility dependence on RIE time. We found that for the film thicknesses of around 900 nm the mobility dropped from 75 to 36 cm^2/(Vs) when the glass surface has been textured by RIE. Sheet resistance of FTO films on RIE substrates are measured to be high, because the film is winding as shown in Fig. 2(b). As the result, low mobility are observed on RIE substrates. However, the low mobility doesn't correspond to the correct property of the film material itself. So, we call it as "Effective mobility". We also found that the film thickness should be much thicker than 2612 nm in order to suppress the drop of the effective mobility. It can be understood that the valley of RIE substrate is filled with film and the film surface becomes flatter. From this result, it is clear that much thicker film than 2612 nm is required to avoid the drop of the effective mobility.

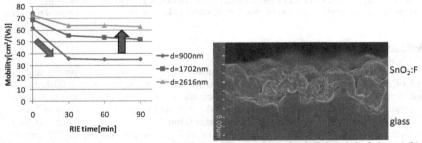

Fig. 2: (a) Effective mobility dependence on RIE time. (Film configuration is F-doped single layer.) (b) SEM cross section image of the sample with d=900nm and RIE time = 30min.

65

Mobility in lower carrier concentration

If film thickness is increased from 900nm to about 3000nm, we need to decrease carrier concentration from $1x10^{20}$ /cm^3 to $2-3x10^{19}$ /cm^3, while keeping free carrier absorption low. Fig. 3 shows the mobility dependence on carrier concentration of FTO films fabricated on flat substrates at various conditions. The mobility of FTO films have the peaks at carrier concentration of about $1x10^{20}$ /cm^3.

At higher carrier concentration, the mobility of FTO films well agree with the theoretical curve considered ionized impurity scattering (eq.(7.7) in Ref[7]), optical phonon scattering (eq. (7.3) in Ref[7]), and acoustic phonon scattering (eq. (7.5) in Ref[7]). All physical properties of SnO$_2$ used in the calculation are shown in Table 7.7 in Ref[7]. At the carrier concentration below $1x10^{20}$ /cm^3 , the mobility drops rapidly from the theoretical curve. This can be due to the grain barrier scattering [7]. Therefore high mobility with low carrier concentration could not be obtained at the same time. In order to solve these constrains, we have adopted a stacked structure using thick non-doped layer and thin F-doped layer. With this approach, the mobility in low carrier concentration is improved as shown in Fig. 3.

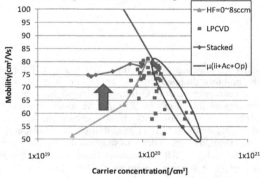

Fig. 3: Mobility dependence on carrier concentration of various samples (HF=0~8sccm: triangle, LPCVD under various conditions : square, Stacked structure : diamond symbol) and theoretical curve (solid line)

Stacked structure on RIE substrate

We adopt this stacked structure to RIE substrate. Fig. 4 shows the mobility and the product of carrier concentration and film thickness (n * d) dependence on film thickness. In stacked structure samples, F-doped and non-doped thickness is varied as n*d value is kept almost same around $7x10^{15}$/cm^2. (F-doped thickness is 200 ~ 600 nm. Non-doped thickness is 1200 ~ 3000 nm.) The mobility of FTO film on RIE substrates is much lower than that on flat substrates if film thickness is thin. If thickness become thick up to 3000 nm, the mobility on RIE and flat substrates are almost the same. According to Fig. 4(b), n * d value is proportional to film thickness for the samples with F-doped single layer , but n * d is kept almost constant in stacked samples, whereas the mobility dependence of these two film structures are almost the same for both flat and RIE substrates.

Table 1 shows measured properties of various samples. SEM image, spectral absorption and haze ratio of these samples are shown in Fig. 5 and Fig. 6. We have successfully achieved the high mobility (80 cm^2/Vs), low carrier concentration (2.2x10^{19} /cm^3) and high haze value (77% at wavelength of 1000 nm) at the same time. This new developed high-haze SnO$_2$ is a new promising TCO for thin-film Si solar cells.

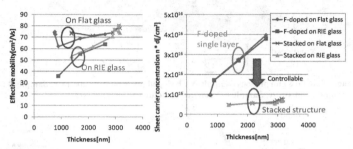

Fig. 4 : (a) Effective mobility dependence and (b) n * d dependence on film thickness. F-doped single layer on flat substrate (diamond), F-doped single layer on RIE substrate (square), stacked structure on flat substrate (cross), stacked structure on RIE substrate.

Table 1: Sample properties of various structure.

	Type-U	Sample A	Sample B	Sample C
Glass substrate	Flat	Flat	Flat	RIE
TCO	F-doped	F-doped	Stacked	Stacked
Thickness[nm]	867	867	2974	3129
Film absorption[%] (ave400 ~ 2000nm)	19.9	6.9	6.2	6.4 (Estimated)
Haze[%] (λ=1000nm)	1.9	0.6	7.9	76.5
Rs[$\Omega/\square$]	6.7	11.0	11.7	11.2
Mobility[cm^2/Vs]	58.8	72.4	76.6	79.8
Carrier concentration [/cm^3]	1.8x10^{20}	9.0 x10^{19}	2.3x10^{19}	2.2x10^{19}
n * d [/cm^2]	1.6x10^{16}	7.8x10^{15}	7.0x10^{15}	7.0x10^{15}

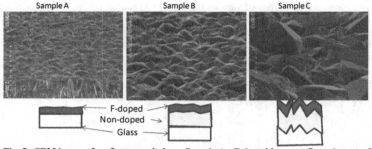

Fig. 5 : SEM image of surface morphology. Sample A : F-doped layer on flat substrate, Sample B : Stacked structure on flat substrate, Sample C : Stacked structure on RIE substrate.

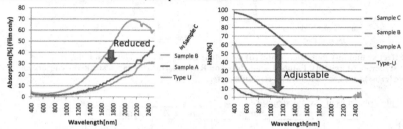

Fig. 6 : (a) Spectral absorption. (b) Spectral haze.

CONCLUSIONS

We have combined F-doped SnO_2 (FTO) with high mobility deposited by LPCVD and reactive ion etching (RIE) processed glass substrate. However, two problems have been found. (1) the mobility of FTO on RIE substrate dropped from that on flat glass (75 to 36 $cm^2/(Vs)$). To avoid this drop, thicker film is needed. (2) To keep high transmittance with thicker film, lower carrier concentration is needed. But the mobility dropped with lower carrier concentration. In order to solve these constrains, we have adopted a stacked structure using thick non-doped layer of 2700 nm and thin F-doped layer of 500 nm. With this novel approach, we have successfully achieved the high mobility (80 $cm^2/(Vs)$), low carrier concentration (2.2×10^{19} /cm^3) and high haze value (77% at wavelength of 1000 nm) at the same time. This new developed high-haze SnO_2 is a new promising TCO for thin-film Si solar cells.

ACKNOWLEDGMENTS

This work was supported by the New Energy and Industrial Technology Development Organization (NEDO) under the Ministry of Economy, Trade, and Industry, Japan (METI).

REFERENCES

1. P. Drude: Z. Phys. **1** (1900) 161 [in German].
2. M. Isshiki, T. Ikeda, J. Okubo, T. Oyama, E. Shidoji, H. Odaka, P. Sichanugrist, and M. Konagai: Jpn. J. Appl. Phys. **51** (2012) 095801.
3. S. Nakao, N. Yamada, T. Hitosugi, Y. Hirose, T. Shimada, and T. Hasegawa: Appl. Phys. Express **3** (2010) 031102.
4. S. Nakao, N. Yamada, T. Hitosugi, Y. Hirose, T. Shimada, and T. Hasegawa: Phys. Status Solidi C **8** (2011) 543.
5. A. Hongsingthong, T. Krajangsang, I. A. Yunaz, S. Miyajima, and M. Konagai: Appl. Phys. Express **3** (2010) 051102.
6. M. Mizuhashi, Y. Gotoh, and K. Adachi: Jpn. J. Appl. Phys. **27** (1988) 2053.
7. K. Ellmer, in Handbook of Transparent Conductors, edited by D. S. Ginley (Springer , New York, 2010), Chap. 7.

Novel Silicon-based Devices and Solar Cells

Mater. Res. Soc. Symp. Proc. Vol. 1536 © 2013 Materials Research Society
DOI: 10.1557/opl.2013.747

Amorphous silicon based betavoltaic devices

N. Wyrsch[1], Y. Riesen[1], A. Franco[1], S. Dunand[1], H. Kind[2], S. Schneider[2], C. Ballif[1]
[1]Ecole Polytechnique Fédérale de Lausanne (EPFL), Institute of Microengineering (IMT),
mb-microtec, Freiburgstrasse 634, 3172 Niederwangen, Switzerland.

ABSTRACT

Hydrogenated amorphous silicon betavoltaic devices are studied both by simulation and experimentally. Devices exhibiting a power density of 0.1 μW/cm^2 upon Tritium exposure were fabricated. However, a significant degradation of the performance is taking place, especially during the first hours of the exposure. The degradation behavior differs from sample to sample as well as from published results in the literature. Comparisons with degradation from beta particles suggest an effect of tritium rather than a creation of defects by beta particles.

INTRODUCTION

Batteries based on radioactive sources, using betavoltaic effect, may offer advantages compared to classical chemical batteries for applications requiring very long lifetime. The betavoltaic principle is based on the direct generation of an electrical voltage and current upon irradiation of a diode by beta particles. Such a principle has been used for power generation in few satellites, but never commercialized. However, betavoltaics has recently seen a renew interest for the powering of micro-devices [1,2].

For safety reasons, it is attractive to use tritium (T) as a low energy beta particle emitter (with 5.7 keV average energy) but it then limits considerably the power density of such a device. Such a low power is compensated by the relatively long T half-lifetime 12.3 years, which should allow for batteries with high energy density. Furthermore, T is industrially available, relatively cheap and can be embedded, if needed, in a solid matrix. Recently, a new battery design has been proposed comprising a 3D porous silicon diode to increase the surface interaction between the active volume (containing T) and the semiconductor device to improve performances [3].

We here explore the possibility to use hydrogenated amorphous silicon (a-Si:H) thin film device together with tritium to fabricate betavoltaic battery. In contrast to the solution incorporating a 3D porous silicon diode design, a-Si:H diodes potentially offer several advantages. a-Si:H cells can be deposited on thin and (eventually) flexible substrates in order to create a 3D module design by stacking or folding such devices. They also exhibit higher performances at low excitation levels compared to c-Si diodes. A few experiments were already performed by other groups using either tritiated amorphous silicon [4] or diodes exposed to tritium [5]. In the latter case, a rapid degradation of the cell was observed, attributed to a replacement of hydrogen in a-Si:H T, leading to an increase of dangling bonds upon decay of tritium atoms into helium [6].

In the present work we will present the potential, the theoretical and practical performance of a-Si-H betavoltaic devices and analyzed the degradation of various a-Si:H diode structures exposed to tritium. Following excellent initial device performance (corresponding to the simulation), a rapid degradation is also observed which differs in some cases from the ones reported in the literature. Comparisons of this degradation from tritium exposure with the one of beta irradiation suggest that T rather than beta particles is involved.

SIMULATIONS

The power density of T at 1 bar is 43.8 µW/cm³. For a betavoltaic device, this power is not directly available. Beta particles emitted from the T should interact with a semi-conductor layer to generate electron-hole (e-h) pairs and the latter must be collected by the device electrodes. Both T and electrodes will also absorb beta particles leading to an attenuation of the beta flux into the device active volume. Furthermore, under the very low excitation given by the beta flux, additional loss has to be expected in the collection of carriers. In order to estimate the power output of the betavoltaic device a comprehensive simulation has been carried out.

We considered the configuration shown in Fig. 1, but neglected all lateral effects; the T layer is assumed to have a much larger spatial extension than the betavoltaic device. Simulations of the beta beam interaction with the a-Si:H based device (as shown in Fig. 2) were done using CASINO software [7]. As this software is restricted to unidirectional particle beam, absorption of betas as a function of depth was computed as a function of the angle of incidence of the incoming beam into the device. This calculation was also carried out as a function of energy as the emission of betas from T follows a distribution of energy with an average of 5.7 keV and a maximum of 18 keV. The flux of betas was then integrated for each angle and energy, taking into account the self-absorption of beta in T from the simplified formula [8] $F = 1 - e^{\mu x}$ where F is the fraction of energy absorbed, μ the linear attenuation factor and x the travelling distance of the particle. For the simulations we used μ=1.81 cm⁻¹ and a T activity of 2.372 Ci/cm³.

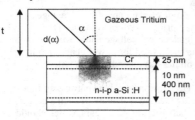

Figure 1. Configuration used for the simulations: An amorphous silicon n-i-p cell (with typically 400 nm thick i-layer and 10 nm thick doped layers) with a thin top contact (here 25 nm thick Cr layer) is put in contact with a layer of T of thickness t. We consider all betas hitting the cell surface coming from the apparent thickness d which depends on the angle of incidence α.

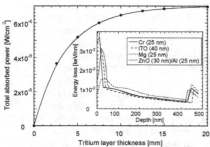

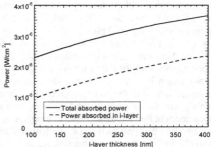

Figure 2. (left) Total power density deposited (in the diode and top contact) as a function T layer thickness in a n-i-p diode with 25 nm thick Cr contact, 10 nm thick doped layer and 400 nm thick i-layer, and (right) the total power density deposited in the diode and the useful power density (deposited in the diode i-layer) as a function of the intrinsic layer thickness. Energy deposited per unit depth for a single beta is shown in the inset for different top contact materials.

Fig. 2 (left) shows the power deposited in the device as a function of the T layer thickness. Due to the self-absorption of the betas by T, the flux and the corresponding energy at the device surface saturate for T thickness above 5 mm. For maximum power collection per unit volume of T, a battery configuration with two betavoltaic devices facing each other and separated by less than 3 mm of T is therefore advisable. The energy deposited as a function of depth for a single beta is also plotted in the inset for several types of top contact material and thickness. In order to minimize the absorption in top contact (which acts as a dead layer), material with low atomic number and low density should be selected. The total and useful power densities are plotted in Fig. 2 (right). The useful power density is given by the power absorbed in the i-layer.

The energy deposited by the betas lead to the generation of e–h pairs with a creation energy of ca. 4 eV (reported values are between 3.4 and 6 eV [9,10]). The useful power of the sample of Fig. 1 corresponds, for full e–h pair collection, to a short-current I_{sc} of 0.58 µA/cm^2. Assuming a minimum fill-factor FF of 25% and an open voltage V_{oc} of 0.5 V (from V_{oc} scaling with generation rate) a total power P of 0.06 µW/cm^2 could be expected. Simulation of the device with ASA software [11], using e–h pair generation profile obtained from above simulations, leads to slightly lower V_{oc} and I_{sc} but larger FF and P (V_{oc}=0.36 V, I_{sc}=0.48 µA/cm^2, FF=58%, P≈0.1 µW/cm^2). Assuming that such thin diode could be deposited on a very thin substrate, with cells facing each others separated by a 2.5 mm layer of T, a power density above 0.5 µW/cm^3 could be achieved (with at least 8 cm^2 of active area per cm^3 of battery).

EXPERIMENT

Test a-Si:H diodes with an area of 0.25 cm^2 were deposited on glass substrate by VHF PECVD (very high frequency plasma enhanced chemical vapor deposition) [12]. Depositions were performed at a plasma excitation frequency of 70 MHz and at deposition temperatures of ca. 200 °C. From the simulation results, two different thicknesses were chosen for the i-layer (250 and 400 nm) with constant thickness (10 nm) for the doped layer. Various top contact materials were tested for the top contact with layer thickness between 20 and 40 nm. These top contacts comprised Al, ITO, ZnO, ZnO/Al, Mg, Al/Mg layers or layer combinations.

Prior to exposing to T, I(V) characteristics of the samples were measured under white light from a sun simulator at various illumination levels using neutral density filters, down to the one corresponding to a comparable e-h pairs generation as in T atmosphere (ca. 0.0045% sun). For characterization in T atmosphere, a dedicated container was built and hooked to a T supply system at mb-microtec. The container was designed in such a way to have a T layer above the sample with a thickness of 2.5 mm. I(V) characteristics were then recorded during several days using a Kethley 2612 sourcemeter. Degradation tests of diodes under electron irradiation were performed in a Philips ESEM XL30 (Environmental Scanning Electron Microscope) with a beam energy of 6 keV. The beam was intentionally defocused in order to minimize the electron-hole pair generation rate and scanned over the diode area in order to obtain uniform degradation. The energy deposited in the i-layer during electron irradiation was estimated to 2.6x10^{17} eV/cm^2 from CASINO simulation, corresponding to ca. 5 hours of T exposure.

RESULTS AND DISCUSSION

Fig. 3 exhibits the I_{sc}, V_{oc} and FF values extracted from I(V) characteristics under white light from a solar simulator, measured on a 400 nm thick diode with various contacts. Ellipses

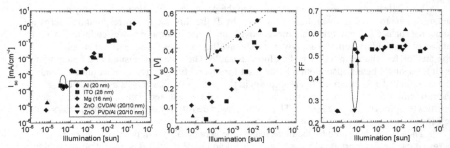

Figure 3. V_{oc}, I_{sc} and FF as a function of light intensity of best 400 nm thick a-Si:H diodes with various top contact. Ellipses indicate target values expected under T exposure (simulations).

indicate the expected values from the simulations. A nice match between experimental and simulated data is obtained for I_{sc}. This fact is not surprising as the equivalent target light intensity was adjusted to obtain the same generation in the diode i-layer as under T exposure. While the change of V_{oc} as a function of light illumination follows the same trend for all samples, the absolute values exhibit a large scattering, especially for contacts deposited by sputtering (ITO and ZnO). This observation seems to be linked to the small (nominal) thickness of theses layers as diode with much thicker contact layer exhibit in general much higher V_{oc} values. Nevertheless, for the best diodes, V_{oc} values close to the target ones were achieved. Mg contact resulted in very poor V_{oc} probably due to the strong degradation of the Mg layer that was observed, when deposited on a-Si:H. Some scattering is also observed in the FF values, especially at low illumination. However, part of the effect can be attributed to the evaluation procedure, as no fitting of the measured I(V) curves were performed, and the FF was deduced from the point with the maximum power. For these relatively small samples measured under very low light illumination a significant noise level induces relatively large errors in the FF determination. However, FF values above 50% were recorded for illumination levels equivalent to T exposure. From these experiments we can conclude that with $V_{oc} \approx 0.35$ V, $I_{sc} \approx 0.4$ μA/cm^2 and FF≈55% a power of ca. 0.08 μW/cm^2 should be attained under T exposure, as given by the simulations.

Several samples were exposed to T and the I(V) characteristics were recorded as a function of time. Fig. 4 shows I_{sc}, V_{oc}, FF and power values of two 400 nm thick diodes with ITO contact (sample A and sample B of Fig, 3) as a function of time. I(V) characteristics of sample A just after the filling of the container with T were the following: I_{sc} = 0.44 μA/cm^2, V_{oc}=0.49 V, FF=42% and P=0.09 μW/cm^2. A large and rapid degradation of I_{sc} as well as of the V_{oc} were observed, while the FF decreased and recovered. A remaining power of 10nW/cm^2 was measured after about 70 h of exposure to T. Sample B exhibits a rather different behavior. Here also I_{sc} and V_{oc} decrease rapidly while FF decreased but did not recovered. Fluctuations in V_{oc} correlate with a variation in temperature during the experiment. Two other diodes with Al or ZnO contact layers exhibited similar behavior as sample B (as of Fig. 4). Samples A and B have a similar structure and top contact but exhibit different degradation kinetics and the origin of these differences and if contacts may influence degradation are still unknown.

The degradation behavior observed above looks at first glance similar to the one reported by Deus et al. [5], who observed a steady degradation of I_{sc} of the diodes exposed to T, as well as

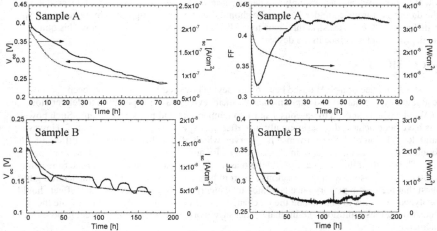

Figure 4. Power, FF, I_{sc} and V_{oc} of two 400 nm thick diodes with ITO top contact layer (sample A – top - and sample B - bottom) as a function of time under T exposure.

a strong initial degradation followed by a stabilization of V_{oc} and FF. Such a behavior, was attributed to an exchange of H by T in the i-layer of the sample and consequently by an increase in deep defect density. However, in our case, the FF of sample A remains at a relatively high level (after the recovery phase). Such behavior, together with the decrease in I_{sc} cannot be explained easily with a simple increase in defect density in the intrinsic layer. Measurement of the I(V) characteristics of the samples after T removal of the container to probe for any T intake by the sample remained inconclusive. In contrast, behavior of sample B, looks more similar to the one of Deus, but no conclusion can be drawn at this stage on the origin for these discrepancies and if the degradation is driven by the diffusion of T into the diode or by defect creation by beta particles.

In order to get more insight into the degradation mechanism, similar samples (to the ones so far studied), were characterized under low light illumination and degraded using the electron beam of the ESEM scanned over the entire surface of the diode. In order to minimize the generation density of electron-hole pairs and to improve uniformity, the beam was defocused and the experimental setup was chosen to present a significant overlap of the line scan with a low scanning speed. The total number of e-h pairs generation during the degradation experiment was adjusted to be the same as under ca. 5 hours of T exposure (biggest changes in the diode performance occurs during the first few hours, just after the beginning of T exposure). The diodes were then again characterized again under low light illumination. A very limited degradation of the performance was observed (in contrast to the ones of Fig. 4): At low light intensity roughly equivalent to T exposure, I_{sc} was observed to decrease from 2.4 to 2.3 µA/cm^2 while FF was reduced from 60 to 55% following the electron beam irradiation

Beta irradiations a-Si:H diodes using SEM microscopes were already performed in the past by two different groups as mean to study defect creation [13,14]. These studies using 20 keV beams show that a significant increase of the defect densities occurs for deposited energy well above 0.1 J/cm^2. In the present study the deposited energy for the beta irradiation was kept below this threshold, and accordingly, no degradation was expected and observed. Degradation of the diodes performance under T exposure should then linked, as suggested by Deus et al. [5], to T diffusion in the device structure. Due to the low H mobility, it is not clear how this quite rapid degradation take place in the devices.

CONCLUSIONS

Performance of a-Si:H based betavoltaic device powered by tritium have been investigated by simulation and experimentally. Such devices have the potential to deliver close to $0.1~\mu W/cm^2$ when exposed to gaseous T at room temperature and pressure of 1 atm. Such power levels have been achieved experimentally just after the exposure of the devices with T. However, a rapid degradation of the performance is observed which affects different samples differently. I_{sc} is found to decrease considerably in all cases while FF and V_{oc} either decline or fluctuate.

The origin of the degradation and the reasons for the discrepancies are for the time being unknown. Comparisons with degradation tests from beta particles coming from an electron gun suggest that the loss in performance is related to T incorporated in the device. Such effect should be analyzed in details to understand the mechanism and find suitable solution to mitigate the effect. Among these solutions, additional barrier layers or new device structures could be implemented.

a-Si:H remains an attractive material for betavoltaic batteries powered by T. The possibility to integrate serially monolithic interconnected device in the battery container would allow for a cost effective design of batteries with customized voltage output. Deposition of the betavoltaic a-Si:H device on thin substrate should enable attractive power density values without relying on 3D diode architecture.

ACKNOWLEDGMENTS

This work was supported by Swiss Commission for Technology and Innovation and by the Swiss National Science Foundation under Contract 200021_126926/1.

REFERENCES

1   L.C: Olsen et al., Physics Today 65 (2012) 35.
2. B. Liu et al., Appl. Phys. Lett. 92 (2008) 083511.
3. W. Sun et al., Adv. Mater. 17 (2005) 1230.
4. T. Kosteski et al., IEE Proc.-Circuits Devices Syst. 150 (2003) 274.
5. S. Deus, Proc. of the 28th Photovoltaic Specialists Conf. (2000) 1246.
6. S. Zukotynski et al., J. of Non-Cryst. 299-302 (2002) 476.
7. http://www.gel.usherbrooke.ca/casino/What.html
8. DOE Handbook, DOE-HDBK-1132-99.
9. L. Hamel et al., IEEE Trans, on Nucl. Sc. 38 (1991) 251.
10. V. Perez-Mendes et al., Nucl. Instr. and Meth. in Phys. Res. A273 (1998) 127.
11. B.E. Pieters et al., Conf. Rec. of the IEEE 4th World Conf. on PV Ener. Conv (2006) 1513.
12. N. Wyrsch et al., MRS Proc. Symp. Vol. 869 (2005) 3-14.
13. U. Schneider et al., J. of Non-Cryst. Sol. 114 (1989) 633
14. F. Diehl et al., J. of Non-Cryst. Sol. 198-200 (1996) 436

Mater. Res. Soc. Symp. Proc. Vol. 1536 © 2013 Materials Research Society
DOI: 10.1557/opl.2013.728

SiC monolithically integrated wavelength selector with 4 channels

M. Vieira[1,2,3], M. A. Vieira[1,2], V. Silva [1,2], P. Louro[1,2], J. Costa [1,2]

[1]Electronics Telecommunication and Computer Dept. ISEL, R. Conselheiro Emídio Navarro, 1959-007 Lisboa, Portugal
[2] CTS-UNINOVA, Quinta da Torre, Monte da Caparica, 2829-516, Caparica, Portugal.
[3] DEE-FCT-UNL, Quinta da Torre, Monte da Caparica, 2829-516, Caparica, Portugal

ABSTRACT

In this paper we present a monolithically integrated wavelength selector based on a double pin/pin a-SiC:H integrated optical active filter that requires optical switches to select visible wavelengths. Red, green, blue and violet pulsed communication channels are transmitted together, each one with a specific bit sequence. The combined optical signal is analyzed by reading out the generated photocurrent, under violet (400 nm) background applied either from the front or the back side of the device. The front and back backgrounds acts as channel selectors that selects one or more channels by splitting portions of the input multi-channel optical signals across the front and the back photodiodes. The transfer characteristics effects due to changes irradiation side are presented. The relationship between the optical inputs and the corresponding digital output levels is established through a 16-element look-up table to perform the optoelectronic conversion.

Results show that the wavelength selector acts as a reconfigurable active filter that enhances the spectral sensitivity in a specific wavelength range and quenched it in the others, tuning a specific band. A binary weighted RGBV code that takes into account the specific weights assigned to each bit position is presented and establishes the optoelectronic functions.

INTRODUCTION

Reconfigurable multi-rate next generation optical networks are currently investigated to handle the ever increasing growth of the Internet traffic. A demultiplexer sends data from a single source to one of several destinations. Whereas the multiplexer is a data selector, the demultiplexer is a data distributor. Reconfigurable wavelength selectors that allow for operation on a large number of wavelength channels, with dynamic response, are essential sub-systems for implementing reconfigurable WDM networks and optical signal processing [1, 2]. This constitutes a solution in WDM technique for information transmission and decoding in the visible range [3].

Amplification and magnitude variation are two key functionality properties outcome of a balanced interaction between the wavelength of the optical signals and background side and wavelength. Any change in any of these factors will result in filter readjustments. Here, signal variations with and without front and back backgrounds move electric field action up and down in a known time frame selecting the appropriated wavelength.

DEVICE CONFIGURATION AND OPERATION

The wavelength selector filter is realized by using double pin/pin a-SiC:H photodetector with front and back biased optical gating elements as depicted in Fig. 1. The filter consists of a p-i'(a-SiC:H)-n/p-i(a-Si:H)-n heterostructure with low conductivity doped layers. The thicknesses and optical gap of the front i'- (200 nm; 2.1 eV) and back i- (1000 nm; 1.8 eV) layers are optimized for light absorption in the blue and red ranges, respectively [4].

Red, green, blue and violet pulsed communication channels ($\lambda_{R,G,B,V}$; input channels) are transmitted together, each one with a specific bit sequence. The combined optical signal (multiplexed signal) is analyzed by reading out the generated photocurrent under negative applied voltage (-8V), without and with violet background (λ=400nm) applied either from the device front or back sides. The device operates within the visible r ange

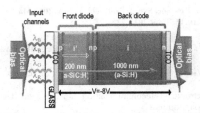

Figure 1. Device configuration and operation.

using as input color channels (data) the wave square modulated light (external regulation of frequency and intensity) supplied by a red (R: 626 nm; 25 $\mu W/cm^2$), a green (G: 524 nm; 46 $\mu W/cm^2$), a blue (B: 470 nm; 40 $\mu W/cm^2$) and a violet (V: 150 $\mu W/cm^2$) LED's. The steady state violet irradiation (optical bias) was superimposed using a LED (400 nm; 2800 $\mu W/cm^2$).

BACKGROUND CONTROLLED SPECTRAL SENSITIVITY

The spectral sensitivity under violet background and without it was tested through spectral response measurements applied either from the front or back side of the device (Fig. 2).

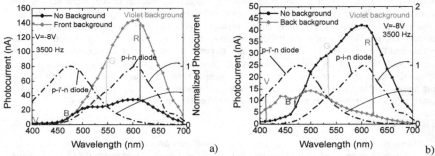

Figure 2. Photocurrent without and with front (a) and back (b) violet background. The normalized photocurrent of the individual photodiodes is superimposed (dash lines).

In Fig. 2a the optical bias was applied from the front side and in Fig. 2b from the back side. For comparison the normalized spectral photocurrent for the front, p-i'-n, and the back, p-i-n, photodiodes (dash lines) are superimposed.

Data shows that the front and back diodes, separately, presents the typical responses of single p-i-n cells with intrinsic layers based on a-SiC:H or a-Si:H materials, respectively. The front diode cuts the wavelengths higher than 550 nm while the back one rejects the ones lower than 500 nm. The overall device presents an enlarged sensitivity when compared with the individual ones.

Under front irradiation the sensitivity is much higher than under back irradiation. Under front irradiation the violet background amplifies the spectral sensitivity mainly in the long wavelength range (>550 nm) while back irradiation strongly quenches this and enhances the short wavelength range (see arrows in both figures). Thus, back irradiation, tunes the front diode while front irradiation selects the back one.

OPTICAL BIAS AMPLIFICATION

To analyze the device under information-modulated wave and uniform irradiation, four monochromatic pulsed lights, separately (red, green, blue and violet input channels; R, G, B, V) illuminated the device at 12000 bps. Steady state violet optical bias was superimposed separately from the front (Fig. 3a) and the back (Fig. 3b) sides and the photocurrent generated measured at -8 V. In both figures the signals, without applied optical bias, are shown. The ratio between the photocurrents with and without background (amplification factor) for the four individual channels ($\alpha_{R\,front}$, $\alpha_{G\,front}$, $\alpha_{B\,front}$, $\alpha_{V\,front}$) are also displayed.

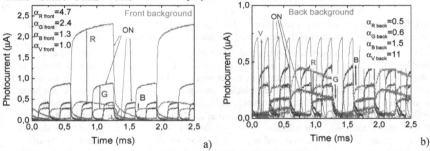

Figure 3. Red (R), green (G) and blue (B) and violet (V) input channels, at -8V, without and under violet steady state optical bias (ON) applied from the front (a) and from the back (b) sides.

Results show that when steady state irradiation is imposed the output photocurrent due to the input signals exhibits a nonlinear dependence on the wavelength. Under front irradiation, the light-to-dark sensitivity is enhanced mainly in the long- medium wavelength ranges. Violet radiation is absorbed at the top of the front diode, increasing the electric field at the least absorbing cell [5], the back diode, where the red and part of the green channels generate optical carriers. So, red and green collection are strongly enhanced (α_{Rfront}=4.7, α_{Gfront}=2.4), the blue lightly increases (α_{Bfront}=1.3) and the violet one stays near its dark value (α_{Bpin1}=1.0). Under back irradiation, the small absorption depth of the violet photons in the back diode enhances the electric field at the front diode and so, the red and green collections are reduced (α_{Rback}=0.5, α_{Gback}=0.6). The violet and the blue channels are absorbed across the front diode resulting in an increase collection for the blue (α_{Bback}=1.5) and enormous increase for the violet (α_{Vback}=11.0) since its absorption occurs near the front p-i′ interface where the electric field suffers the larger increase.

DATA SELECTOR

In Fig. 4a the combined signal (multiplexed signal) is displayed with front and back violet background. In Fig. 4b the difference between the multiplexed signals under front and back irradiation, the wavelength-generation rate, is depicted. On the top the signals used to drive the input channels are shown to guide the eyes into the ON/OFF states of each input. In Fig. 4b, for comparison the signal under front irradiation is added.

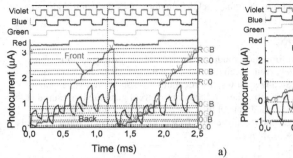 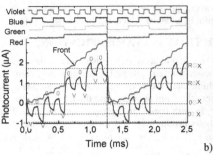

Figure 4. a) Multiplexed output signal under front and back violet irradiation. b) Wavelength difference generation. On the top the signal to drive the input channels guide the eyes.

Under front irradiation, data shows that the encoded signal (output) presents sixteen ordered levels (2^4 inputs, dotted lines) each one related with an *RGBV* bit sequence (right side of Fig. 4a). The signal magnitude is balanced by the channel amplification factor (Fig. 3a, $\alpha_{RGBV,front}$) of each input. Depending on the presence or lack of an input signal, respectively 1 or 0 in a binary code, and on its amplification factor, the levels can be grouped into subgroups. The highest amplification of the red channel allows the division of the 16 inputs into two groups of 8 entries each: the upper 8 ascribed to the presence of the red channel (R=1) and the lower 8 to its absence (R=0). The green channel presents a medium amplification, so, the four (2^2) highest levels, of each group of 8, are ascribed to the presence of the green channel (G=1) and the four lower ones to its lack (G=0). The blue channel is slightly amplified, so, in each group of 4 entries, two groups (2^1) can be found: the two higher levels correspond to the presence of the blue channel (B=1) and the

two lowers to its absence (B=0). Finally, each group of 2 entries have two sublevels, the higher where the violet channels is ON (V=1) and the lower where it is missing (V=0). The selection index for this 16-element look-up table is a 4-bit binary *R,G,B,V* code of the form S_3, S_2, S_1,S_0 where *Sn* means the color channel (Fig. 4a) with *n* proportional to the amplification factor (Fig. 3a). So, the multiplexer select code, under front irradiation, represents an address (RGBV), or index, into the ordered inputs. Under back irradiation the amplification factors change in an opposite way (Fig. 3b), the violet and blue channels are enhanced and the green and red reduced. The high amplification of the violet channel allows the division of the 16 inputs into two main groups of 8 entries each. The uppers 8 ascribed to the presence of the violet channel (V=1) and the lower 8 to its absence (V=0) acting as a 8-to-1 multiplexer.

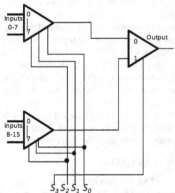

Figure 5. Schematic of a pin/pin a-SiC:H 16-element table look-up.

In each group the 4 higher levels have the blue channel ON and the others OFF (4-to-1 multiplexer). The upper level has both red and the green channels ON and the lower both OFF, in the other two middle entries only one of these channels is present. So, the 4-bit binary multiplexer select code represents a *VBGR* address into the ordered input. We may view the 16-element look-up table as consisting of two look-ups, one to select the proper group of 8, and pick the red under front irradiation (S_3=R) or the violet under back background (S_3=V). Each group of 8 inputs requires 3 bits for picking the proper group of 8 and for specifying an input (S_2, S_1, S_0). In Fig. 5 the schematic of the pin/pin a-SiC:H 16-element table look-up is displayed. Here, if S_3=R then the S_2=G, S_1=B and S_0=V; if S_3=V the 4-bit binary code will be BGR.

DATA ROUTER

Whereas the multiplexer is a data selector, the demultiplexer is a data distributor or data router. Just as the multiplexer has a binary code (RGBV) for the selection of an input, the demultiplexer has a similar code for selecting a particular output. In the pi'n/pin device the side of the background is the routing control for the data source. The front and back background acts as selector to select one of the four incoming channels by splitting portions of the input multi-channel optical signal across the front and back photodiodes (Fig 2, see arrows). This duality of functions is characteristic of decoders and demultiplexers.

Under front background the red channel is decoded due to its higher amplification while under back violet irradiation the violet channel is selected (Fig. 4a). To help to decode the green and blue channels, in Fig. 4b the difference between the multiplex signal under front and back violet irradiation is displayed. This difference wavelength generation is a consequence of nonlinear interaction of the device with the front or back backgrounds and the optical channels generation. It weights the red versus violet content of the measured signal, so, it enhances the effect of the routing control and offers a transparent wavelength conversion. The presence of the red channel pushes the difference up and the violet channel pushes it down (right side of Fig. 4b). The blue channel does not affect the difference. So, after decoding the red and the violet transmitted information and comparing with difference wavelength generation levels in the same time slots, the green and blue signals can be immediately decoded. We have used this simple algorithm to perform 1 -to-16 demultiplexer (DEMUX) function and to decode the multiplex signals. As proof of concept the decoding algorithm was implemented in *Matlab* [6] and tested using different binary sequences. In Fig. 6a a random MUX signal under front and back irradiation is displayed and in Fig. 6b the decoding results are shown. On the top of both figures the signals to drive the LED's and the DEMUX signals obtained as well as the binary bit sequences are respectively displayed. A good agreement between the signals used to drive the LED's and the decoded sequences is achieved. In all sequences tested the RGBV signals were correctly decoded.

The DEMUX sends the input logic signal to one of its sixteen outputs, according to the optoelectronic demux algorithm. So, by means of optical control applied to the front or back diodes, the photonic function is modified, respectively from a long- pass filter to pick the red channel to a short-pass filter to select the violet channel, giving a step reconfiguration of the device. The green and blue channels are selected by combining both active long- and short-pass filters into a band-pass filter. In practice, the decoding applications far outnumber those of demultiplexing. Multilayer SiC/Si optical technology can provide a smart solution to communication problem by providing a possibility of optical bypass for the transit traffic by dropping the fractional traffic that is needed at a particular point.

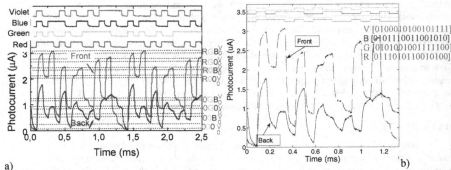

Figure 6. MUX signal under front and back irradiation. On the top a) Signals used to drive the LED's. b) DEMUX signals and decoded RGBV binary bit sequences.

CONCLUSIONS

A monolithically integrated wavelength selector based on SiC multilayer photonics active filters is analyzed. Tailoring the filter wavelength was achieved by changing the violet background side. Results show that the pin/pi´n multilayered structure becomes reconfigurable under front and back irradiation, acting either as data selector or data router. It performs WDM optoelectronic logic functions providing photonic functions such as signal amplification, filtering and switching.

A 16-element look-up table was presented. The selection index is a 4-bit binary RGBV code under front irradiation and VBGR under back light. An algorithm to decode the four input channels is presented.

ACKNOWLEDGEMENTS

This work was supported by FCT (CTS multi annual funding) through the PIDDAC Program funds and PTDC/EEA-ELC/111854/2009 and PTDC/EEA-ELC/120539/2010.

REFERENCES

1. P.P. Yupapin, P. Chunpang,", Int. J. Light Electron. Opt., 120(18) (2009) 976-979.
2. S. Ibrahim, L. W. Luo, S. S. Djordjevic, C. B. Poitras, I. Zhou, N. K. Fontaine, B. Guan, Z. Ding, K. Okamoto, M. Lipson, and S. J. B. Yoo, paper OWJ5. Optical Fiber Communications Conference, OSA/OFC/NFOEC, San Diego, 21 Mar 2010.
3. S. Mitatha, K. Dejhan, P.P. Yupapin and N. Pornsuwancharoen., Int. J. Light Electron. Opt. (2008). doi:10.1016/j.ijleo.2009.03.012.
4. M.A. Vieira, P. Louro, M. Vieira, A. Fantoni, A. Steiger-Garção. IEEE sensor jornal, Vol. 12, NO. 6, (2012) pp. 1755-1762.
5. M. Vieira, A. Fantoni, P. Louro, M. Fernandes, R. Schwarz, G. Lavareda, and C. N. Carvalho, Vacuum, 82, Issue 12, 8 August 2008, pp: 1512-1516.
6. M. A. Vieira, M. Vieira, J. Costa, P. Louro, M. Fernandes, A. Fantoni, in Sensors & Transducers Journal Vol. 10, Special Issue, February 2011, pp.96-120.

Mater. Res. Soc. Symp. Proc. Vol. 1536 © 2013 Materials Research Society
DOI: 10.1557/opl.2013.754

Design of an optical transmission WDM link using plastic optical fibers

P. Louro[1,2], P. Soares[1], H. Ferraz[1], P. Pinho[1,4], M. Vieira[1,2,3,]
[1] Electronics, Telecommunications and Computer Dept., ISEL, Lisbon, Portugal
[2] CTS-UNINOVA, Lisbon, Portugal
[3] DEE-FCT-UNL, Quinta da Torre, Monte da Caparica, 2829-516, Caparica, Portugal
[4] Instituto de Telecomunicações, Aveiro, Portugal

ABSTRACT

In this paper we present the design of an optical transmission system, using plastic optical fiber (POF), which operates in the visible range of the electromagnetic spectrum. The optical signals are generated by modulated visible LEDs, transmitted through POF and at the reception end a pin-pin photodetector is implemented. A computer simulation tool dedicated to the analysis of optical circuits was used for preliminary analysis of the optical system. The performance of the optical link was analyzed by BER prediction variation on the transmission rate. The tested optical system was assembled using high efficiency LEDs of the same wavelengths, a commercial POF and a pin-pin photodetector based on a-SiC:H/a-SI:H. This detector behaves as an optical filter with controlled wavelength sensitivity. Different optical signals, obtained by adequate modulation of LED optical sources, were coupled into the POF and the combined optical signal at the fiber termination was directed onto the photodetector active area. The output photocurrent was measured with and without optical bias. Results compare the use of a pin-pin transducer device in free space and in a POF transmission link.

INTRODUCTION

Traditionally, optical communications use infrared windows of the electromagnetic spectrum, due to the low attenuation provided by glass optical fibers in these wavelengths. The development of other types of fibers, namely, plastic optical fibers (POF), allowed the possibility of using other regions of the spectrum. Actually, in the visible range, this technology constitutes a good alternative for broadband communications over short distances (up to about 1 km), successfully replacing coaxial and twisted-pair cables or even multimode glass optical fiber [1, 2]. Their main advantages are light weight, great flexibility, operation in the visible range and easier coupling with the optical source (large numerical aperture). This type of transmission systems are used for short range communication links, e. g., automotive vehicles, domestic and office LANs, sensors, lighting, medicine instruments, etc.. The capacity of an optical communication system can be increased using the Wavelength Division Multiplexing (WDM) technique [3], where several optical signals are combined and transmitted together through the same medium.

We propose a communication system operating in the visible range using three ultra-bright LEDs of different wavelengths, a POF and an integrated photodetector and demultiplexer device based on two stacked multilayered a-SiC:H/a-Si:H structures that act as optical filters in the visible range [4]. The possibility of tuning the spectral device sensitivity is analyzed and discussed using several optical bias conditions that induce different modulations of the electrical field along both front and back heterostructures, amplifying or cutting specific wavelengths [5]. This enables the identification of the transmitted individual input channels and implements the demultiplexing operation in the visible spectrum.

PARAMETERIZATION OF THE SYSTEM

A simple optical communication system to transmit visible signals over POF was designed (Fig. 1). The simulations of the system were made using the VPItransmissionMaker™ 8.5 software [6].

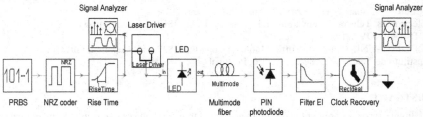

Fig. 1 System for transmission of visible optical signals over POF.

The different modules of the simulated optical system are: *PRBS* (generates a pseudorandom binary sequence that represents the information to be transmitted), *NRZ coder* (performs a digital to electrical conversion using a NRZ coding), *Rise Time* (gaussian filter used to smooth the electrical signal), *Laser Driver* (drives electrical signals into the optical sources), *LED* (represents the optical source), *Multimode fiber* (represents a POF using the proper parameters), *PIN Photodiode* (photodetector module), *Filter El* (universal filter used as a low-pass filter), *Clock Recovery* (synchronizes the received electrical signal with the original signal) and *Signal Analyzers* (to compare the transmitted and the received signal).

Optical Source

The parameterization of the optical sources included three different visible emission wavelengths for LED light: 470 nm (blue), 525 nm (green) and 625 nm (red). The emission spectral model was defined as flat.

Optical Fiber

For the optical fiber two graded index POFs were chosen, namely, OM-Giga with a polymethylmethacrylate (PMMA) core and Lucina that uses a perfluorinated polymer core (CYTOP® from Asahi Glass Company). Notice that the datasheets of these fibers don't present the attenuation values for the used wavelengths. So, the attenuation values for 650 nm operation were used in the simulations. The parameterization of the optical fibers is shown in Table I.

TABLE I. OPTICAL FIBERS PARAMETERS [1]

Parameter	OM-Giga	Lucina
Material	PMMA	CYTOP®
Wavelength (nm)	470, 525 and 625	
Attenuation (dB/km)	200	40
Core Refractive Index	1. 49	1.34
Numerical Aperture	0.3	0.195
Core Diameter (µm)	900	120

Photodetector

PIN photodiodes were used to simulate the behavior of the heterostructure used in this work as integrated photodetector and WDM device. The responsivity was parameterized using experimental specific data values of the device at the wavelengths of interest: 0.1 A/W at 470 nm, 0.5 A/W at 525 nm and 1.1 A/W at 625 nm. The dark current was assumed 1 nA and the thermal noise 10 pA/Hz$^{1/2}$.

EVALUATION OF THE SYSTEM

In order to compare the role of different types of optical sources on the communication system performance we simulated the bit error rate (BER) variation with the transmission rate, considering a 100 m link of POF. Results are displayed in Fig. 2. The value of BER = 10^{-9}, considered usually as the lowest limit for a reliable data transmission system [1], is also shown as a reference quality value. BER limits of 10^{-12} and 10^{-6} were assumed in the analysis.

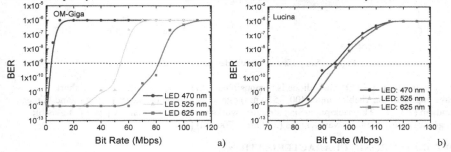

Fig. 2 LEDs performance on a 100 m link of POF: a) OM-Giga and b) Lucina.

Results show that with the OM-Giga POF (Fig. 2a), the LED wavelength determined the system performance. Better results were obtained with the red LED (625 nm), with transmission rates up to 80 Mbps. With the green LED (525 nm) reliable transmission was obtained only up to 50 Mbps, and for the blue light (470 nm) the performance was very poor. This behavior results from the huge number of propagated modes in this fiber (8.1×10^5 at 470 nm, 6.5×10^5 at 525 nm and 4.6×10^5 at 625 nm), due to the large dimensions of the core (900 µm). This effect induces great intermodal dispersion and consequently increases the BER, which leads to huge signal degradation. With the Lucina POF (Fig. 2b) the BER variation obtained with different LED wavelengths has a similar trend, which means that with this fiber the wavelength of the LED, at least in this range, is not a crucial parameter on the optical system performance. As can be seen, the best transmission rates were of about 95 Mbps. Its smaller core (120 µm) significantly reduced the number of propagated modes (6.1×10^3 at 470 nm, 4.9×10^3 at 525 nm and 3.5×10^3 at 625 nm), and thus there was less intermodal dispersion and better performance than with the OM-Giga POF.

It was also observed that the achieved transmission rates were of the order of megabits per second. This is explained because LEDs are much more limited than other types of optical sources such as, for example, laser diodes. If the optical source was a distributed feedback (DFB) laser or a vertical-cavity surface-emitting laser (VCSEL), the achievable bit rates could be of the order of gigabits per second. However, the results obtained in the simulations show that this type of fibers is appropriated to be used in the next experimental scenarios.

PHOTODETECTOR DEVICE

Fig. 3 shows the simplified cross-section structure used as integrated photodetector and demultiplexer device. It is a multilayer heterostructure composed by two pin structures built on a glass structure and sandwiched between two transparent electrical contacts.

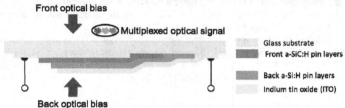

Fig. 3 Simplified schematic diagram of the device structure.

The front pin a-SiC:H photodiode is responsible for the device sensitivity in the short wavelengths of the visible range (400 – 550 nm) due to its narrow thickness (200 nm) and higher bandgap (2.1 eV). The back pin a-Si:H structure works in the complimentary past of the visible range, collecting the long wavelengths (520 nm – 700 nm) [7].

OPTOELECTRONIC CHARACTERIZATION

Spectral sensitivity

Fig. 4 displays the photocurrent, measured along the visible spectrum, under reverse bias without and with optical light bias focusing the device from back and front sides. Results of Fig. 4 show that the use of steady state light bias induces changes in the spectral sensitivity of the device. Front violet bias enhances the signals of long wavelengths (> 500 nm), while the violet bias imposed from the back side causes the opposite effect, as output photocurrent signal increase is observed in the short wavelengths range.

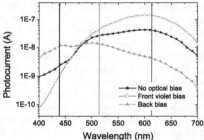

Fig. 4 Spectral photocurrent under dark conditions and using front and back violet light.

Transient input signals

The analysis of the transient device response used three monochromatic waveform optical signals driven by ultra-bright LEDS (Red, Green and Blue input channels) that illuminated the device from the front side (Fig. 3). The optical signals were transmitted through a PMMA core

POF, and the signal at the reception end was directed onto the device sensitive area (front side). Optical bias was also used to soak uniformly the device separately from the front and back sides of the device. In Fig. 5 the transient signals of each individual optical channel without and with violet front and back illumination are plotted. On the top of the figure it is displayed the optical signal. As anticipated from spectral response data (Fig. 5) the device sensitivity is strongly dependent on the optical bias. For long wavelengths (green and red) it is observed amplification on the photocurrent under front violet optical bias while under back bias the signal is reduced. This effect is more evident for longer wavelengths, as the amplification factor for the red channel is higher than in the green.

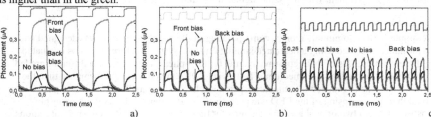

Fig. 5 a) Red, b) green and c) blue transient input signals measured without optical bias and under violet front and back optical bias.

On the other hand, short wavelength signals (blue) are amplified under back violet background and reduced with front violet light. Thus the use of a short front wavelength for the optical bias (400 nm) will strongly amplify the red channel, produce an increase in the green channel and make no effect on the blue channel. On the other hand, if the violet light is used from the back side this will result in amplification of the blue channel and attenuation of the red and green channels.

Transient combined signal

In order to test the transmission of multiple optical signals through the POF combined red, green and blue optical waveform signals were used. Fig. 6 displays the transient multiplexed signal travelling in free space and through the POF using different optical bias conditions during the signal acquisition process: no optical bias (dark line), and front (violet line) and back violet light (blue line) for the steady state light.

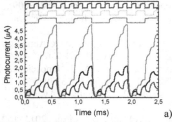 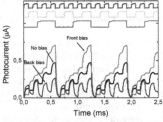

Fig. 6 Combined signals acquired without and with front and blue back optical bias transmitted through: a) free space and b) POF. On the top the input optical signals are shown.

In Fig. 6a the output signal measured without optical bias exhibits 8 photocurrent thresholds, each one assigned to the different ON-OFF states of the input optical signals. The highest level corresponds to the situation where all input channels are ON, while the lowest to the condition of OFF. Under front bias it is observed a strong amplification when the red channel is ON, while under back bias it is the blue channel that becomes enhanced. In Fig. 6b) the output magnitude signal is very low. This loss usually assigned to attenuation phenomena is probably due to the poor coupling between the optical sources and the POF, which resulted into a reduced performance. However, the shape of the output photocurrents measured under different optical bias are in shape agreement with the same signals measured in free space.

CONCLUSIONS

A simple optical communication system operating in the visible spectrum was designed and its performance analyzed using a dedicated simulation tool. System performance was evaluated using three visible LEDs of different wavelengths for the optical source and two different graded index POFs for the transmission medium. The Lucina POF showed better performance than the OM-Giga POF, due to reduced intermodal dispersion related to its smaller core. It was also observed that the increasing of the LED wavelength allowed better results. A double pin-pin heterostructure with two optical gate connections for light triggering in different spectral regions was used. BER variation on the transmission rate was evaluated to infer the system performance. Spectral response characterization of the photodetector under different optical biasing conditions showed the dependence on the device absorption mechanism. Multiple monochromatic pulsed signals, in the visible range, were transmitted together, each one with a specific bit sequence, through a POF link and in free space. Future work comprises the enhancement of the coupling between the POF and the optical sources and the adequate parameterization of the pin-pin detector.

ACKNOWLEDGEMENTS

This work was supported by FCT (CTS multi annual funding) through the PIDDAC Program funds and PTDC/EEA-ELC/120539/2010.

REFERENCES

1 O. Ziemann, J. Krauser, P. E. Zamzow e D. Werner, POF Handbook Optical Short Range Transmission Systems, 2nd Edition, Springer-Verlag Berlin Heidelberg, 2008.
2 S. Louvros and I. E. Kougias, Contemporary Eng. Sciences, vol. 2, n.° 2, pp. 47-58, 2009.
3 M.Bas, Fiber Optics Handbook, Fiber, Dev.and Syst. for Opt. Comm., Chap, 13, Mc Graw-Hill, 2002.
4 P. Louro, M. Vieira, M. A. Vieira, S. Amaral, J. Costa, M. Fernandes, Sensors & Actuators: A Physical vol. 172 (1), 35-39 (2011) http://dx.doi.org/10.1016/j.sna.2011.01.026..
5 Louro, P.; Vieira, M.; Fernandes, M.; Vieira, M. A.; Costa, J.; Fantoni, A., J. Nanoscience and Nanotechnology 11(6) 2011, pp. 5318-5322(5). DOI: 10.1166/jnn.2011.3779.
6 VPIphotonics™, "VPItransmissionMaker™ Optical Systems," 2012. [Online]. Available: http://www.vpiphotonics.com.
7 P. Louro, M. Fernandes, J. Costa, M. A Vieira, A. Fantoni, M Vieira, MRS Proc. 2011, 1321, 223-228.

Mater. Res. Soc. Symp. Proc. Vol. 1536 © 2013 Materials Research Society
DOI: 10.1557/opl.2013.703

Optoelectronic logic functions using optical bias controlled SiC multilayer devices

M. A. Vieira[1,2], M. Vieira[1,2,3], V. Silva[1,2], P. Louro[1,2], M. Barata[1,2]

[1]Electronics Telecommunication and Computer Dept. ISEL, R. Conselheiro Emídio Navarro, 1949-014 Lisboa, Portugal Tel: +351 21 8317290, Fax: +351 21 8317114, mv@isel.ipl.pt ;
[2] CTS-UNINOVA, Quinta da Torre, Monte da Caparica, 2829-516, Caparica, Portugal.
[3] DEE-FCT-UNL, Quinta da Torre, Monte da Caparica, 2829-516, Caparica, Portugal

ABSTRACT

The purpose of this paper is the design of simple combinational optoelectronic circuit based on SiC technology, able to act simultaneously as a 4-bit binary encoder or a binary decoder in a 4-to-16 line configurations. The 4-bit binary encoder takes all the data inputs, one by one, and converts them to a single encoded output. The binary decoder decodes a binary input pattern to a decimal output code.

The optoelectronic circuit is realized using a a-SiC:H double pin/pin photodetector with two front and back optical gates activated trough steady state violet background. Four red, green, blue and violet input channels impinge on the device at different bit sequences allowing 16 possible inputs. The device selects, through the violet background, one of the sixteen possible input logic signals and sends it to the output.

Results show that the device acts as a reconfigurable active filter and allows optical switching and optoelectronic logic functions development. A relationship between the optical inputs and the corresponding digital output levels is established. A binary color weighted code that takes into account the specific weights assigned to each bit position establish the optoelectronic functions. A truth table of an encoder that performs 16-to-1 multiplexer (MUX) function is presented.

INTRODUCTION

There has been much research on semiconductor devices as elements for optical communication, when a band or frequency needs to be filtered from a wider range of mixed signals or when optical active filter are used to select and filter input signals to specific output ports in WDM communication systems [1, 2].

Optical communication in the visible spectrum usually interfaces with an optoelectric device for further signal processing. Multilayered structures based on amorphous silicon technology are expected to become reconfigurable to perform WDM optoelectronic logic functions [3, 4]. They will be a solution in WDM technique for information transmission and decoding in the visible range [5]. The basic operating principle is the exploitation of the physical properties of a nonlinear element to perform a logic function, with the potential to be rapidly biasing tuned Any change in any of these factors will result in filter readjustments. Here, signal variations with and without front and back backgrounds move electric field action up and down in a known time frame. A truth table support new optoelectronic logic architecture.

DEVICE OPERATION

The optoelectronic circuit consists of a p-i'(a-SiC:H)-n/p-i(a-Si:H)-n heterostructure with low conductivity doped layers as displayed in Fig.1. The optoelectronic characterization was described elsewhere [6]. Monochromatic pulsed lights, separately (λ_R=626 nm, λ_G=526 nm, λ_B=470 nm, λ_V=400 nm; input channels) or in a polychromatic mixture (multiplexed signal) at different bit rates illuminated the device.

Independent tuning of each channel is performed by steady state violet optical bias (λ_{bias}= 2300 µW/cm^2) superimposed either from the front and back sides and the generated photocurrent measured at -8V. The device operates within the visible range using as input color channels (data) the wave square modulated light (external regulation of frequency and intensity) supplied by a red (R; 25 µW/cm^2), a green (G; 46 µW/cm^2), a blue (B; 40 µW/cm^2) and violet (V; 150 µW/cm^2) LED's.

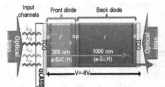

Figure 1. Device configuration and operation.

OPTICAL BIAS CONTROLLED FILTERS

In Fig.2, the spectral photocurrent, normalized to its value without background is displayed, under front (a) and back (b) violet irradiations and different intensities.

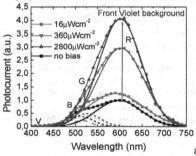
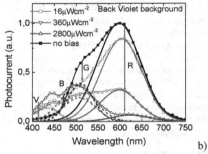

a)   b)

Figure 2. Normalized spectral photocurrent under front (a) and back (b) violet irradiations with different intensities.

A peak fit adjustment to the data was performed (lines) with peaks centered on 630 nm (solid), 520 nm (dash) and 430 nm (dot). Results show that under front violet irradiation, as the background intensity increases, the peak centered at 630 nm (red range) strongly increases while under back light an opposite behavior is observed and the red peak is strongly reduced (see arrows). Under front and back side irradiation, the peak at 520 nm (green range) increases slightly with the intensity. Under back irradiation, a new peak centered at 430 nm appears and increases with the background intensity. So, under front illumination the reddish part of the spectrum is strongly enhanced with the intensity while under back illumination the main enhancement occurs at the violet-blue region. A trade-off between the background intensity and the enhancement or quenching of the different spectral regions, under front and back irradiation, has to be established.

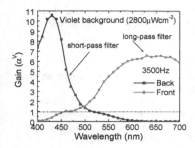

Figure 3. Spectral gain under violet (α^V) optical bias, applied from the front and the back sides at 3500Hz.

In Fig. 3 the spectral gain (α^V), defined as the ratio between the spectral photocurrents under violet illumination (applied from the front and back sides) and without it, is plotted at 3500 Hz and 2300μWcm^{-2}. As expected from Fig. 2, under back bias the gain is high at short wavelengths and strongly lowers for wavelengths higher than 500 nm, acting as a short-pass filter. Under violet front light the device works as a long-pass filter for wavelengths higher than 550 nm, blocking the shorter wavelengths. Results show that by combining the background wavelengths and the irradiation side the short-, and long- spectral region can be sequentially tuned. The medium region can only be tuned by using both active filters.

ENCODER AND DECODER DEVICE

Optical switching

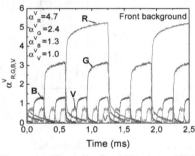

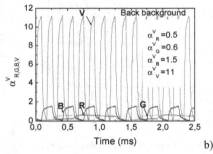

Figure 4. Normalized red (R), green (G) blue (B)and violet (V) transient signals at -8V with violet (400 nm) steady state optical bias applied from the front (a) and from the back (b) sides.

Four monochromatic pulsed lights separately (red, green, blue and violet input channels, Fig. 4) or combined (multiplexed signal, Fig. 5) illuminated the device at 12000 bps. Steady state violet optical bias was superimposed separately from the front (a) and the back (b) sides and the photocurrent measured. In Fig. 4, the transient signals were normalized to their values without background and added the mean values of the optical gains for each individual channel.

Results show that, even under transient conditions and using commercial LED's as pulsed light sources, the background side affects the signal magnitude of the color channels. As in Fig. 2, under front irradiation, it enhances mainly the spectral sensitivity in the medium-long wavelength ranges (α^V_R=4.7, α^V_G=2.4). Violet radiation is absorbed at the top of the front diode, increasing the electric field at the back diode [7] where the red and part of the green incoming photons are absorbed (see Fig. 1). Under back irradiation the electric field increases mainly near the front p-n interface where the violet and part of the blue incoming channels generate most of the photocarriers ($\alpha^V_V=11$, $\alpha^V_B=1.5$). So, by switching between fronts to back irradiation the photonic function is modified from a long- to a short-pass filter allowing, alternately selecting the red or the violet channels.

Optoelectronic logic functions

For an optoelectronic digital capture system, opto-electronic conversion is the relationship between the optical inputs and the corresponding digital output levels.

Fig. 5 displays the normalized MUX signals due to the combination of the input channels of Fig. 4, without and under front (a) and back (b) violet irradiations. On the top the signals used to drive the input channels are displayed showing the presence of all the possible 2^4 *on/off* states. For comparison the MUX signal without optical bias is displayed (dark) in both figures.

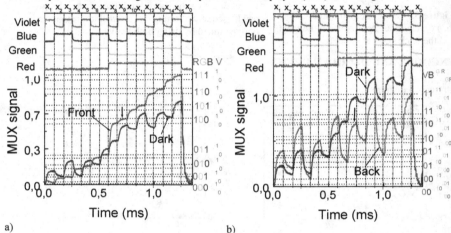

a)   b)

Figure 5. Normalized multiplexed signal under front (a) and back (b) violet irradiation and without it (dark). On the top the signals used to drive the input channels are shown to guide the eyes into the ON/OFF channel states.

Results show that the side of background affects the form and the magnitude of the MUX signal in opposite ways. Under front irradiation, sixteen levels (2^4) are detected and grouped into two main classes due to the high amplification of the red channel ($\alpha^V_R \gg 1$; Fig. 4a). The upper eight (2^3) levels are ascribed to the presence of the red channel (R=1), and the lower eight to its absence (R=0), allowing the red channel decoder (8-to-1 multiplexer; long-pass filter function). Since under front irradiation the green channel is also amplified,($\alpha^V_G > 1$) the four (2^4) highest levels, in both classes, are ascribed to the presence of the green channel (G=1) and the four lower ones to its lack (G=0). The blue channel is slightly amplified, so, in each group of 4 entries, two levels (2^1) can be found: the two higher levels correspond to the presence of the blue channel (B=1) and the two lowers to its absence (B=0). Finally, each group of 2 entries have two near sublevels, the higher where the violet channels is ON (V=1) and the lower where it is missing (V=0).

Under back irradiation, the violet channel is strongly enhanced, the blue channel is slightly and the green and red reduced (α^V_R and $\alpha^V_G<1$ and $\alpha^V_B>1$ and $\alpha^V_V>>1$; Fig. 4b). The encoded multiplexed signal is, also, made of sixteen sublevels grouped into two main levels, the higher where the violet channel is ON (V=1) and the lower where it is OFF (V=0) (8-to-1 multiplexer; short-pass filter function). Each group the eight sublevels can be grouped in two classes, with and without the blue channel ON. Each of those classes split into four near sublevels, attributed to the presence or absence of the red and green channel. If we consider this red and green output bits "not significant" only four separate levels (2^2) are considered and the logic MUX function is converted into a logic filter function. The violet channel is then decoded.

MUX
⇐===========⇒
DEMUX

Inputs   **Outputs**   Y

x_1	x_2	x_3	x_4	x_5	x_6	x_7	x_8	x_9	x_{10}	x_{11}	x_{12}	x_{13}	x_{14}	x_{15}	x_0	S_3	S_2	S_1	S_0
0	0	0	0	0	0	0	0	0	0	0	0	0	0	1	0	1	1	1	1
0	0	0	0	0	0	0	0	0	0	0	0	0	1	0	0	1	1	1	0
0	0	0	0	0	0	0	0	0	0	0	0	1	0	0	0	1	1	0	1
0	0	0	0	0	0	0	0	0	0	0	1	0	0	0	0	1	1	0	0
0	0	0	0	0	0	0	0	0	0	1	0	0	0	0	0	1	0	1	1
0	0	0	0	0	0	0	0	0	1	0	0	0	0	0	0	1	0	1	0
0	0	0	0	0	0	0	0	1	0	0	0	0	0	0	0	1	0	0	1
0	0	0	0	0	0	0	1	0	0	0	0	0	0	0	0	1	0	0	0
0	0	0	0	0	0	1	0	0	0	0	0	0	0	0	0	0	1	1	1
0	0	0	0	0	1	0	0	0	0	0	0	0	0	0	0	0	1	1	0
0	0	0	0	1	0	0	0	0	0	0	0	0	0	0	0	0	1	0	1
0	0	0	1	0	0	0	0	0	0	0	0	0	0	0	0	0	1	0	0
0	0	1	0	0	0	0	0	0	0	0	0	0	0	0	0	0	0	1	1
0	1	0	0	0	0	0	0	0	0	0	0	0	0	0	0	0	0	1	0
1	0	0	0	0	0	0	0	0	0	0	0	0	0	0	0	0	0	0	1
0	0	0	0	0	0	0	0	0	0	0	0	0	0	0	1	0	0	0	0

Figure 6. Truth table of the encoders that perform 16-to-1 multiplexer (MUX) function, under front violet irradiations.

The binary code is an arithmetic code and so, it is weighted, *i. e.* there is specific weights assigned to each bit position. Due to the different optical gains (Fig. 3), the selection index for those 16-element look-up table are a 4-bit binary [RGBV] code under front irradiation or a 4-bit binary [VBGR] of the form [S_3, S_2, S_1, S_0] where S_n means the color channel (right side of Fig. 5) with *n* weighted by the amplification factor (Fig. 4). The multiplexer select code represents an address or index, into the ordered inputs.

The truth tables of the encoder of Fig. 5a, that perform 16-to-1 MUX function, is shown in Fig. 6. The correspondence between the on/off state of the input channels and the [RGBV] code are obvious. In the inputs (x_0....x_{15}), the index of each bit, is related to the first (highest) nonzero logic input. Here, the MUX device selects, through the front violet background, one of the sixteen possible input logic signals and sends it to the output ($y=x_S$). The output is a 4-bit binary RGBV number that may identify one of sixteen possible inputs. Just as the multiplexer has a binary code for the selection of an input, the demultiplexer (DEMUX) has a similar code for selecting a particular output. The 4-bit output RGBV code allows designing an encoder to transform a four-line-to-sixteen-line decoder. From truth table of Fig. 6, the Boolean functions for the encoder with inputs x_0 to x_{15} and outputs R, G, B, V is given as:

$R(S_3)=\sum(8,9,10,11,12,13,14,15)$; $G(S_2)=\sum(4,5,6,7,12,13,14,15)$;
$B(S_1)=\sum(2,3,6,7,10,11,14,15)$; $V(S_0)=\sum(1,3,5,7,9,11,13,15)$.

A binary representation for decimal number 9 is in RGBV code "1001" (2^3+2^0) under front irradiation and it corresponds to both red and violet channels ON. Under back irradiation (VBGR code) the binary representation is the same although the weights assigned to each bit position are different (see arrows in Fig. 5). This 4-bit output RGBV code allows us design a 4-to-10 line decoder to transform a decimal number (0 to 9) into a binary code. The 4-bit codes from 1010 through 1111 do not arise from the encoding of the decimal numbers.

CONCLUSIONS

An optoelectronic device based on a-SiC:H technology is analyzed. The device is able to act simultaneously as a 4-bit binary encoder or a binary decoder in a 4-to-16 line configurations.

A relationship between the optical inputs and the corresponding digital output levels is established. A binary weighted color code that takes into account the specific weights assigned to each bit position establish the optoelectronic functions. A truth table of an encoder that performs 16-to-1 multiplexer (MUX) function is presented. More work as to be done in order to execute optical arithmetic micro-operations entirely within the optical domain.

ACKNOWLEDGEMENTS

This work was supported by FCT (CTS multi annual funding) through the PIDDAC Program funds and PTDC/EEA-ELC/111854/2009 and PTDC/EEA-ELC/120539/2010.

REFERENCES

1. C. Petit, M. Blaser, Workshop on Optical Components for Broadband Communication , ed. by Pierre-Yves Fonjallaz, Thomas P. Pearsall, Proc. of SPIE Vol. 6350, 63500I, (2006).
2. S. Ibrahim, L. W. Luo, S. S. Djordjevic, C. B. Poitras, I. Zhou, N. K. Fontaine, B. Guan, Z. Ding, K. Okamoto, M. Lipson, and S. J. B. Yoo, paper OWJ5. Optical Fiber Communications Conference, OSA/OFC/NFOEC, San Diego, 21 Mar 2010.
3. M.A. Vieira, P. Louro, M. Vieira, A. Fantoni, A. Steiger-Garção. IEEE sensor jornal, Vol. 12, NO. 6, (2012) pp. 1755-1762.
4. M.A. Vieira, M. Vieira, P. Louro, V. Silva, A., Applied Surface Science, DOI: 10.1016/j.apsusc.2013.01.020.
5. S. Randel, A.M.J. Koonen, S.C.J. Lee, F. Breyer, M. Garcia Larrode, J. Yang, A. Ng'Oma, G.J Rijckenberg, and H.P.A. Boom.. "ECOC 07 (Th 4.1.4). (pp. 1-4). Berlin, Germany, 2007.
6. M. Vieira, P. Louro, M. Fernandes, M. A. Vieira, A. Fantoni and J. Costa InTech, Chap.19, pp:403-425 (2011).
7. M. Vieira, A. Fantoni, P. Louro, M. Fernandes, R. Schwarz, G. Lavareda, and C. N. Carvalho, Vacuum, 82, Issue 12, 8 August 2008, pp: 1512-1516.

Mater. Res. Soc. Symp. Proc. Vol. 1536 © 2013 Materials Research Society
DOI: 10.1557/opl.2013.748

Improvement of seed layer smoothness for epitaxial growth on porous silicon

Martini R. [1,2], Sivaramakrishnan Radhakrishnan H. [1,2], Depauw V. [2], Van Nieuwenhuysen K. [2], Gordon I. [2], Gonzalez M. [2], Poortmans J. [1,2]

[1] KU Leuven - Department of Electrical Engineering, Kasteelpark Arenberg 10, 3001 Leuven, Belgium
[2] Imec, Kapeldreef 75, 3001 Leuven, Belgium

ABSTRACT

In the last decades many techniques have been proposed to manufacture thin (<50μm) silicon solar cells. The main issues in manufacturing thin solar cells are the unavailability of a reliable method to produce thin silicon foils with contained material losses (kerf-losses) and the difficulties in handling and processing such fragile foils. A way to solve both issues is to grow an epitaxial foil on top of a weak sintered porous silicon layer. The porous silicon layer is formed by electrochemical etching on a thick silicon substrate and then annealed to close the top surface. This surface is employed as seed layer for the epitaxial growth of a silicon layer which can be partially processed while attached on the substrate that provides mechanical support. Afterward, the foil can be bonded on glass, detached and further processed at module level. The efficiency of the final solar cell will depend on the quality of the epitaxial layer which, in turn, depends on the seed layer smoothness.

Several parameters can be adjusted to change the morphology and, hence, the properties of the porous layer, both in the porous silicon formation and the succeeding thermal treatment. This work focuses on the effect of the parameters that control the porous silicon formation on the structure of the porous silicon layer after annealing and, more specifically, on the roughness of the top surface. The reported analysis shows how the roughness of the seed layer can be reduced to improve the quality of the epitaxial growth.

INTRODUCTION

One of the challenges in today's photovoltaic industry and research is to reduce the thickness of silicon solar cells from 180μm to 40-50μm. The driving force that leads in this direction is two-fold: on one hand, the cost of bare silicon still accounts for most of the cost of a solar module and, on the other hand, simulations show that the maximum efficiency can be reached for such thin substrates [1].

Two main issues arise in processing such thin silicon solar cells; firstly, no reliable method to produce this kind of substrates without material losses exists and, secondly, the fragility of thin silicon substrates makes them very difficult to handle and process.

One of the most promising methods to solve both issues at the same time is the layer transfer technique based on a weak porous silicon (PS) layer. Within this technique a silicon substrate is electrochemically etched to create a stack of a porous silicon with low porosity (LPS) layer on top of a highly porous silicon (HPS) layer which is afterward annealed at high temperature. During the high temperature process, the columnar pores reorganize into sphere-like pores while the top surface closes. This surface is then employed as seed layer for the epitaxial growth of a

40µm-thick silicon layer which can then be detached and processed as a solar cell on low-cost substrates or stand-alone, while the parent wafer is reused [2].

High efficiency silicon solar cells produced with this technique rely on high quality epitaxial growth. Donolato [3] linked the decrease of dislocation density to the increase of minority-carrier lifetime while Alberi et al. [4] showed how the open circuit voltage in epitaxial silicon solar cell increases by decreasing the dislocation density. Recently, Haase et al. [5] presented an analysis of the efficiency of back-contacted solar cells with thicknesses from 290 to 45µm and they reported a drop in efficiency due to a low carrier lifetime in the epitaxially grown base. Since crystalline defects inside epitaxial layers are known to increase with the seed layer roughness, an analysis of the seed layer smoothness prior the epitaxial growth is of foremost importance to enhance the efficiency of epitaxial solar cells.

This work reports on the reduction of the roughness of the PS top surface to improve the quality of the epitaxial layer grown on it. The parameters that can be tuned in the anodization process have been studied, i.e. anodization time, anodization current, electrolyte composition, and their effect on the smoothness of the seed layer have been investigated after annealing with the same conditions. The study of these parameters will help the optimization of the PS structure in order to increase the quality of the epitaxial foil and to reduce, in this way, bulk recombination due to crystalline defects.

EXPERIMENTS

PS double layers have been formed by electrochemical etching of 700µm-thick p-type CZ silicon wafers with resistivity of 0.009Ωcm in anodic regime. Starting from a reference anodization process, the parameters that control the formation of the LPS layer have been modified individually. Three different values have been studied for each parameters and the actual values are reported in table I. Current density and time for the formation of the HPS layer have been kept constant.

Table I. Analyzed values for each parameter. The reference configuration is indicated with bold characters.

Time [s]	Current density [mA/cm^2]	HF concentration [%]
50 – 150 – **300**	0.32 – **1.35** – 6.5	**21** – 27 – 33

After anodization, the samples have been annealed in an epitaxial reactor in H_2 atmosphere at 1 atm for 10 minutes at 1130°C. Once the samples have been annealed, the structure of the PS layers has been analyzed by cross-sectional scanning electron microscopy (SEM) while the roughness of the PS top surfaces has been evaluated by high resolution profilometry on one hundred 20µm-long lines with 5nm-resolution.

RESULTS

By electrochemical etching, a meso-porous silicon layer is formed. During high temperature annealing, the atoms diffuse along the surface in order to minimize the surface energy and, hence, the pores firstly close and, then, transform into spheres-like pores which surfaces are mainly defined by facets with lower surface energy, i.e. {111}, {100} and {311} planes [6]. The equilibrium condition is attained when all the pores show this structure. The morphology of the top surface of the LPS layer which will act as seed layer for epitaxial growth depends on the reorganization process which, in turn, depends on the initial configuration of the PS layer. If one of the tunable parameters is changed, different aspects of the PS structure are affected, e.g. pores dimension, thickness, porosity. Their effect on the PS microstructure and the surface smoothness are hereby discussed.

Anodization current

The anodization current density can be increased to etch larger pores in the silicon substrate. At the same time also the porosity and the etching rate increase. The roughness of the seed layer after reorganization for the three current densities analyzed in the study is reported in Figure 1. The seed layer shows a drastic increase of the surface roughness and a larger spread on the data when higher current densities are applied.

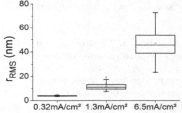

Figure 1. Seed layer roughness for samples with LPS layers anodized by applying different current densities.

The cross-sections of the three sample anodized by changing the current density are depicted in figure 2. By decreasing the current density (left picture), smaller and more faceted pores are obtained after the identical annealing process, hence indicating a faster reorganization.

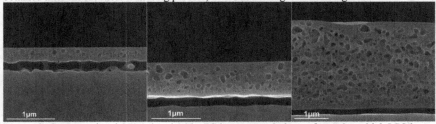

Figure 2. Cross-sectional SEM showing the PS layers morphology of samples which LPS has been anodized for $0.32mA/cm^2$ (left), $1.3mA/cm^2$ (center) and $6.5mA/cm^2$ (right)

Electrolyte composition

In the range of conditions analyzed in this study, the fluorine ions can easily diffuse inside the pores thanks to presence of ethanol in the electrolyte also for higher HF concentrations and, thus, an increase of the HF concentration increases the concentration of fluorine ions at the pores tip, which leads to the formation of smaller pores. A higher HF concentration in the electrolyte solution has also an effect on the etching rate, which is accelerated, and the porosity, which instead decreases. In figure 3 the roughness of the seed layers obtained with different electrolytes is reported. This graph highlights an improvement of the seed layer smoothness when PS is formed with electrolytes that present a higher HF concentration.

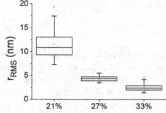

Figure 3. Seed layer roughness for samples that have been etched with electrolytes with different HF concentration.

Figure 4 shows cross-sectional SEM pictures of the porous silicon double layer electrochemically etched by employing electrolyte with different HF concentrations after annealing. These pictures show that, by increasing the HF concentration, smaller and more faceted pores are formed after annealing. Therefore, as in the case lower current densities are applied, it indicates that smoother surfaces are obtained in case LPS layers closer to the equilibrium condition are attained before epitaxial growth. The SEM images also indicate that, due to the decreased porosity and the faster etching rate in case of higher HF concentration, the conditions applied to etch the HPS layer do not allow the formation of a weak layer necessary for the detachment. Therefore, the HPS layer will have to be optimized in case electrolytes with different HF concentration are used.

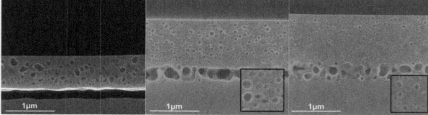

Figure 4. Cross-sectional SEM pictures showing the PS layers morphology of samples anodized in electrolyte solution with 21% (left), 27% (center) and 33% (right) HF concentration. The boxes in the center and left figures show a close-up view of the pores structure

Anodization time

Anodization time has mainly affecting the thickness of the LPS layer. Within the set of analyzed values, the electrolyte can easily diffuse inside the pores and, hence, the etching rate is nearly constant. The anodization time has also a second minor effect on the pores dimension and the porosity due to a longer etching time of the side walls of the pores in contact with the electrolyte [7]. Therefore, with longer anodization times, slightly larger pores and higher porosity are expected.

Figure 5 depicts the top surface roughness of the three sample anodized for different etching times. This graph exhibits a reduction of the roughness of about one order of magnitude when the anodization time is reduced from 300s to 50s and, at the same time, the distribution of the data is narrowed in case of very short anodization time.

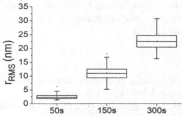

Figure 5. Seed layer roughness for samples with LPS layers that have been etched for different times.

Cross-sectional SEM pictures of the three samples reported in figure 6 display the differences between the pores structures. In the very thin LPS layer, the pores are well aligned and they exhibit a faceted structure which corresponds to a fully reorganized PS layer while, in case of thicker LPS layers, the pores show more irregular shapes.

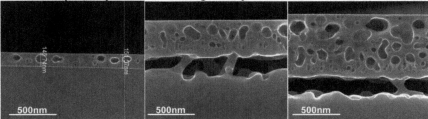

Figure 6. Cross-sectional SEM showing the PS layers morphology of samples which LPS layer has been anodized for 50s (left), 150s (center) and 300s (right)

DISCUSSION

This work has been focused on the effect of each parameter that can be controlled during electrochemical etching on the top surface smoothness and the LPS layer structure. The results reported in the previous section show that the smoothness of the top surface improves when lower current densities are applied, when electrolytes with higher HF concentration are employed and when the anodization time is reduced. Although changing only one parameter at the time can simultaneously affect different characteristics of the porous layer, e.g. thickness, pores dimension, pores density, important considerations can be extracted from these results about the reorganization process.

The roughness measurement and the cross-sectional SEM images manifest an improvement of the top surface roughness in the samples that, after the annealing stage, exhibit pores with a more facet structure, which indicates a faster reorganization process. A faster reorganization can be explained by a reduction of the initial pores dimension in case of lower current densities and higher HF concentrations. In case of shorter anodization time, smaller pores are also obtained because of the shorter time the pore walls are in contact with the electrolyte. However, the etching rate of the pores walls is usually very small, in the order of few angstroms per hour, and therefore it seems to be unlikely to cause a relevant increase of the pores dimension. A further analysis of the pores structure before reorganization is needed in this case to better understand the faster reorganization.

CONCLUSIONS

High efficiency solar cells produced with porous-silicon-based layer transfer techniques rely on high quality epitaxial growth and, hence, on a smooth seed layer. The roughness of such layer is determined by the anodization and annealing conditions. This work has been focused on the evaluation of the roughness of the top surface of samples electrochemically etched in different conditions that underwent, afterward, the same annealing step. The effect of the parameters involved in this process has been analyzed by changing them individually and measuring the roughness of each sample by high resolution profilometry. Three trends have been shown leading, after the annealing step, to smoother surfaces: the shortening of the anodization time, the decrease of the current density and employing electrolytes with higher HF concentration. In these conditions, the pores structure reorganizes faster and smaller pores are obtained, leading to structures closer to the equilibrium after the same annealing conditions. Provided the optimization of the HPS layer in order to obtain a weak layer that allows the detachment of the epitaxial foil from the substrate, these conditions can be applied in the PS-based layer transfer technique to grow higher quality epitaxial layers.

ACKNOWLEDGMENTS

The authors want to thank the agency for innovation in science and technology IWT for financial support within the SiLaSol project (project no. IWT90047).

REFERENCES

1. M. J. Kerr, A. Cuevas and P. Campbell, *Prog. Photovolt: Res. Appl.* **11**, 97-104 (2003)
2. R. Brendel, *Jpn. J. Appl. Phys.* **40**,4431 (2001)
3. C. Donolato, *J.Appl.Phys.* **84**, 2656 (1998)
4. K. Alberi, H. M. Branz, H. Guthrey, M. J. Romero, I. T. Martin, C. W. Teplin, P. Stradins and D. L. Young, *Appl.Phys.Lett.* **101**, 123510 (2012)
5. F. Haase, R. Winter, S. Kajari-Schroder, M. Mese and R. Brendel, *Proceedings of the 27th EU-PVSEC conference* (2012)
6. D. J. Eaglesham, A. E. White, L. C. Feldman, N. Moriya and D. C. Jacobson, *Phys. Rev. Lett.* **70**, 1643-1646 (1993)
7. X. G. Zhang, *Electrochemistry of Silicon and Its Oxide*, (Kluwer Academic/Plenum Publishers, New York, 2001), p.367

Materials and Devices Characterization and Simulation

Mater. Res. Soc. Symp. Proc. Vol. 1536 © 2013 Materials Research Society
DOI: 10.1557/opl.2013.817

Local junction voltages and radiative ideality factors of a-Si:H solar modules determined by electroluminescence imaging

T. M. H. Tran [1], B. E. Pieters[1], M. Schneemann[1], T.C.M. Müller[1], A. Gerber[1], T. Kirchartz[2], and U. Rau[1]

[1]IEK5-Photovoltaik, Forschungszentrum Jülich GmbH, 52425 Jülich, Germany,
email: t.tran@fz-juelich.de
[2]Department of Physics and Centre for Plastic Electronics, Imperial College London,
London SW7 2AZ, United Kingdom

ABSTRACT

In this contribution, we show that the dominant electroluminescent emission of hydrogenated amorphous silicon (a-Si:H) thin-film solar cells follows a diode law, whose radiative ideality factor n_r is larger than one. This is in contrast to crystalline silicon and Cu(In, Ga)Se$_2$ solar cells for which n_r equals one. As a consequence, the existing quantitative analysis for the extraction of the local junction voltage $V_j(r)$ from luminescence images fails for a-Si:H solar cells. We expand the existing analysis method, and include the radiative ideality factor n_r into the model. With this modification, we are able to determine the local junction voltage $V_j(r)$ for a-Si:H solar cells and modules. We investigated the local junction voltage $V_j(r)$ and the radiative ideality factor n_r for both initial and stabilized a-Si:H solar modules. Furthermore, we show that the apparent radiative ideality factor is affected by the spectral sensitivity of the used camera system.

INTRODUCTION

Recently, electroluminescence (EL) has received much attention as a fast inline characterization tool for photovoltaic devices [1-3]. Spatially resolved EL imaging is used to derive minority carrier diffusion length [2], local recombination current, and local series resistance of crystalline silicon solar cells and modules [4,5]. Furthermore, EL is widely used for the detection and quantitative analysis of local defects (shunts) in crystalline silicon solar cells and modules [6,7]. Moreover, Helbig *et al.* applied an EL imaging method to quantitatively analyze local junction voltage differences, sheet resistances of the front- and back contact, and power losses due to series resistances and shunt resistances in Cu(In,Ga)Se$_2$ (CIGS) thin-film solar modules [8]. The determination of local voltage differences is based on the assumption that the emission follows a diode law with a radiative ideality factor of unity, i.e, $n_r = 1$. If we furthermore assume the superposition principle holds [9], absolute values for the junction voltage V_j can be obtained by calibrating the relative local voltage ΔV_j calculated from EL images with the open circuit voltage (V_{oc}). Note that the absolute junction voltage does not include the voltage drop over (internal) series resistances. This method can also be applied directly for crystalline silicon solar cells and modules. However, in case of hydrogenated amorphous silicon (a-Si:H) solar modules, the interpretation of EL images is not as straight-forward because n_r is larger than one, and the superposition principle is not held. In this work, we demonstrate that with a modification of the existing EL analysis method, the radiative ideality factor n_r, and consequently the local junction voltages V_j of a-Si:H solar cells and modules can be quantitatively determined.

THEORY

EL analysis for solar cells with $n_r=1$ (CIGS, c-Si)

The term electroluminescence of a photovoltaic device describes the emission of light by applying a forward voltage bias. It is the reverse action to the standard operation of a solar cell. Due to the validity of the detailed balance theory between absorption and emission, Rau [10] shows that the reciprocity theorem is correct if the radiative ideality factor equals one. The reciprocity theorem allows the determination of the local junction voltage $V_j(r)$ directly from the local EL intensity $\phi_{el}(r,E)$ via

$$\phi_{el}(r,E) = Q_e(E)\phi_{bb}(E)\exp\left(\frac{qV_j(r)}{kT}\right). \tag{1}$$

Here, $Q_e(r,E)$ is the external quantum efficiency, $\phi_{bb}(E)$ is the spectral photon density of a black body, E is the photon energy, and kT/q is the thermal voltage. For a spatially resolved EL analysis, the local dependency of Q_e is often neglected due to the much stronger, exponential impact of the spatial variations of $V_j(r)$ [8].

The EL signal $S_{ccd}(r,E)$ is detected with a charge coupled device camera (CCD). The output of each camera pixel is the integral of the product of the EL intensity $\phi_{el}(r,E)$ and the energy-dependent sensitivity $Q_{CCD}(E)$ of the camera,

$$S_{ccd}(r,E) = \int Q_{ccd}(E)Q_e(E)\phi_{bb}(E)e^{\frac{qV_j(r)}{kT}}\,dE. \tag{2}$$

The local voltage $V_j(r)$ can directly be derived from Eq.((2)

$$V_j(r) = \frac{kT}{q}\ln[S_{ccd}(r,E)] + V_{off}. \tag{3}$$

We name the unknown part of Eq.(3) as an offset voltage V_{off}

$$V_{off} = -\frac{kT}{q}\ln\left[\int Q_{ccd}(E)Q_e(E)\phi_{bb}(E)\,dE\right]. \tag{4}$$

From an EL measurement, we first obtain the relative local junction voltage $\Delta V_j(r)$ via

$$\Delta V_j(r) = \frac{kT}{q}\ln[S_{ccd}(r,E)]. \tag{5}$$

Once, the offset voltage V_{off} is known, the absolute local voltage $V_j(r)$ can easily calculated from Eq.(3).

$$V_j(r) = \Delta V_j(r) + V_{off}. \tag{6}$$

If the superposition principle is valid and the emission follows a diode law with a radiative ideality factor of unity, $n_r = 1$, only a single short circuite current density/open circuit voltage J_{sc}/V_{oc} pair is needed to determine the offset voltage V_{off} [8].

EL analysis for solar cells with $n_r \neq 1$ (a-Si:H)

For a-Si:H solar cells, the requirements for the EL analysis described above are not fulfilled. R.S. Crandall has shown that for a-Si:H devices superposition does not hold [11]. Furthermore, it is well established that EL of a-Si:H devices originates from radiative transition between exponential distributions of localized band tail states [12]. From hydrogenated microcrystalline (μc-Si:H) solar cells, it is known that such radiative tail-to-tail recombination leads to a radiative ideality factor larger than one, and luminescence spectra that change with the applied voltage [13,14]. Thus, we expected that these effects also play a role in a-Si:H devices. For this reason we rewrite Eq.(1) to include a radiative ideality factor n_r

$$\phi_{el}(r, E) = Q_e(r, E)\phi_{bb}(E)\exp\left[\frac{qV_j(r)}{n_r kT}\right].$$ (7)

For the junction voltage, we write

$$V_j(r) = \frac{n_r kT}{q}\ln[S_{ccd}(r, E)] + V_{off}.$$ (8)

We need at least two J_{sc}/V_{oc} pairs measured at different illumination intensities to determine n_r and the V_{off}. It follows

$$n_r = \frac{q}{kT}\left\{\frac{V_{oc}(J_{sc}=J_2)-V_{oc}(J_{sc}=J_1)}{\log[\phi_{el}(J_2)]-\log[\phi_{el}(J_1)]}\right\}.$$ (9)

The absolute local junction voltage V_j is then determined by Eq. (8).

EXPERIMENTAL

The samples used for this work are single junction a-Si:H solar modules (8×8 cm^2). Each module consists of eight series connected cells with respective areas of 8×1 cm^2. Our standard a-Si:H device consists of a Glass/ZnO/a-Si:H p-i-n/ZnO/Ag layer stack. Preparation details can be found in reference [14]. The module is monolithically series connected using three laser scribing processes as described in more detail in reference [15]. Of these devices the illuminated and dark characteristics, i.e, the J/V curves, as well as the short circuit current density J_{sc} and the open circuit voltage V_{oc} of the a-Si:H samples were measured with a class A sun simulator at standard test conditions (STC). Furthermore, the electroluminescence images were taken either with a full frame Si-CCD-camera (9.4 mega pixels) from Apogee imaging systems and an InGaAs-CCD from Princeton Instruments systems with a resolution of 640×512 pixels.

RESULTS AND DISCUSSION

Determination of the absolute local junction voltage from EL images measured by Si-CCD

The determination of the absolute local junction voltage $V_j(r)$ requires the acquisition of several EL images of the same cell at different constant injection current densities. Figures 1 (a) and (b) exhibit EL images of an a-Si:H solar module (8×8 cm^2) at $J = 1.25$ mAcm^{-2} and $J = 15$ mAcm^{-2}, respectively.

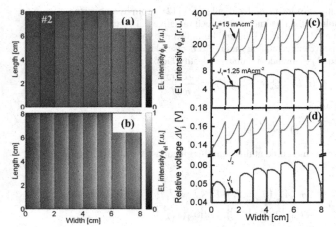

Figure 1: EL images of an a-Si:H module (8×8 cm^2) recorded at different injection current densities: (a) $J = 1.25$ mAcm^{-2}, (b) $J = 15$ mAcm^{-2}. (c) EL line-scans taken from EL images at different injection current densities. (d) Line-scans of the relative local voltages calculated from EL line-scans in (c).

The EL images show an almost defect-free module. There is only one minor shunt in cell #2, which slightly lowers the EL intensity of this cell as observed in Fig.1 (a). At higher injection current density, the EL intensity is much stronger due to exponential relationship between the EL intensity and the local junction voltage. Therefore, the influence of the shunts on the performance of the module becomes negligible. However, the influence of the series resistance becomes more pronounced, which is represented in the decrease of the EL intensity from the right to the left of a cell [see Fig. 1 (b)]. Line-scans shown in Fig.1 (c) are obtained from averaging the EL intensities over the whole length of the module. Line-scans of the relative local voltage are calculated from line-scans of the EL intensity using Eq. (7) after determining the radiative ideality factor n_r from Eq. (9). The line-scan of the relative local voltage at low injection condition ($J = 1.25$ mAcm^{-2}) shows a slightly lower voltage of cell # 2 compared to other cells. The influence of the minor shunt in cell # 2 becomes pronounced at low injection current. However, at high injection current density, the difference between the relative junction voltage ΔV_j at the end and at the beginning of each individual cell becomes larger because of the impact of the sheet resistance of the ZnO front contact.

The relative junction voltage $\Delta V_{j,i}$ of each individual cell i is determined by averaging the relative local junction voltage over the area of the cell. Therefore, the relative junction voltage of the whole module ΔV_j directly results from the sum of $\Delta V_{j,i}$. The absolute junction voltage V_j of the module is calculated from Eq. (8).

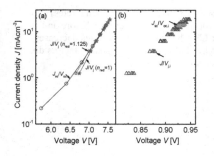

Figure 2: (a) J/V - characteristics of module: J_{sc}/V_{oc} measured with a sun simulator at various irradiation intensities (line with open-circles), J/V_j from EL measurements in case (n_r = 1) (line with full-stars), and (n_r = 1.125) (line with open-stars), respectively. (b) J/V - characteristics of each individual cell: $J/V_{j,i}$ of each cell from EL measurements (n_r = 1.125) (up-triangles), $J_{sc}/V_{oc,i}$ of each individual cell at ($J_{sc} = J$) (full-circles).

Figure 2 (a) shows the J/V characteristics of the module depicted in Fig 1 (a). The J_{sc}/V_{oc} characteristic of the investigated module (line with open-circles) is used to determine n_r [see Eq. (10)] and V_{off}. The J/V_j curve of the whole module (line with open stars) determined from Eq.(8) fits perfectly to the J_{sc}/V_{oc} curve with a radiative ideality factor of $n_r = 1.13$. However, the J/V_j curve (line with full-stars) in case of $n_r = 1$ does not fit to the J_{sc}/V_{oc} curve satisfyingly.

Figure 2 (b) illustrates the J/V characteristics of each individual cell calculated from EL measurements (up-triangles). The deviation of the junction voltage of each individual cell $V_{j,i}$ at low injection current densities is quite small, indicating that the investigated module is almost defect-free. Moreover, the open circuit voltages $V_{oc,i}$ (full-circles) of each individual cell at ($J_{sc} = J$) fit quite well to $V_{j,i}$ at the corresponding current density J. The results show that with an extension of n_r is larger than one to the existing analysis method, both junction voltages of the whole module and of each individual cell can be quantitatively determined.

Influence of light induced degradation of the luminescent properties of the a-Si:H modules

In this part, we investigate the influence of light induced degradation of the luminescent properties of the a-Si:H modules. Therefore, the investigated sample was light-soaked for 60 h at 50°C. The radiative ideality factors n_r and the absolute junction voltage were then determined for both initial and light soaked states.

	J_{sc} [mA/cm^2]	V_{oc} [V]	FF[%]	η [%]	n_r
initial	13.9	7.4	68.3	8.81	1.13
LS	13.1	7.2	57.9	6.85	1.07

Table 1: The module parameters before and after 60 h of light soaking (LS).

The reductions in efficiency η (-22 %) and fill factor FF (-15 %) shown in Tab. 1 after light soaking (LS) are due to the well-known Staebler-Wronski Effect (SWE) in a-Si:H [16],[17]. The open circuit voltage V_{oc} reduces -3 % after LS, which is also shown in the J_{sc}/V_{oc} curves (line with open circles) and J/V_j curves (open stars) before and after LS in Fig. 3. However, the radiative ideality factors obtained from experiments before ($n_r = 1.13$) and after (n_r = 1.07) LS do not change. Note that the values for n_r before and after LS shown in Tab.1 is calculated from EL images recorded by a Si-CCD camera, which means that this result is only valid for its respective spectral range.

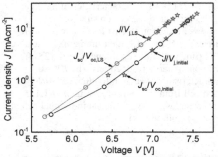

Figure 3: J_{sc}/V_{oc} (line with open circles) and J/V_j (open stars) characteristics before and after LS. The open circuit voltage decreases after LS, but the radiative ideality factors do not change.

Impact of the spectral sensitivities of different camera systems on radiative ideality factor

The reasons why the quantitative EL interpretation applying for CIGS and c-Si, does not work for a-Si:H solar cell are the dependency of its EL spectra on the applied voltage [18], and the radiative ideality factor being larger than one. Figure 4 shows EL spectra of a-Si:H device measured at room temperature ($T = 300$ K) and at different injection current densities J. The higher the injection current density, the more shifts the EL peak to higher photon energies E. Figure 4 also ascribes that the sensitivity of the InGaAs detector is high for the whole spectrum of the a-Si:H solar cell, whereas the spectral sensitivity of the Si detector is very low at lower photon energies. Therefore, n_r found for the investigated a-Si:H sample measured by Si and InGaAs cameras have different values, $n_r = 1.07$ and $n_r = 1.35$, respectively.

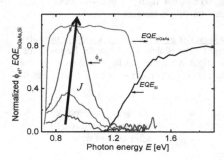

Figure 4: EL spectra of a-Si:H device at different injection current densities and external quantum efficiency (EQE) of Si and InGaAs detectors.

CONCLUSIONS

We present a modification of an existing EL analysis method by including a radiative ideality factor n_r, which is larger than one into this existing model. As a consequence, the absolute junction voltages of each individual cell and of the whole a-Si:H module are quantitatively determined. Thereby, we are able to calculate the radiative ideality factor n_r from EL- and J_{sc}/V_{oc} measurements. The values of n_r calculated from EL images recorded by the Si detector before and after LS do not change. Moreover, using different camera systems with respective spectral sensitivity, yields different values for n_r.

ACKNOWLEDGMENTS

The authors would like to thank J. Kirchhoff, U. Gerhards and C. Zahren for providing the samples and T.C.M. Müller for fruitful discussion about EL spectra of a-Si:H device. This work has been supported by the German Ministry for the Environment, Nature Conservation and Nuclear Safety under the contract FK0325364 and the NRWZiel2 Project "InnoPV" (AZ64.65.69.EN1022A).

REFERENCES

1. O. Breitenstein, J. Bauer, K. Bothe, D. Hinken, J. Muller, W. Kwapil, M. C. Schubert, and W. Warta, *37th IEEE Photovoltaic Specialists Conference* 1, 159 (2011).
2. T. Fuyuki and A. Kitiyanan, *Applied Physics A* 96, 189 (2008).
3. T. Trupke, E. Pink, R. a. Bardos, and M. D. Abbott, *Applied Physics Letters* 90, 093506 (2007).
4. K. Ramspeck, K. Bothe, D. Hinken, B. Fischer, J. Schmidt, and R. Brendel, *Applied Physics Letters* 90, 1 (2007).
5. J. Haunschild, M. Glatthaar, M. Kasemann, S. Rein, and E. R. Weber, *physica status solidi (RRL) - Rapid Research Letters* 3, 227 (2009).
6. O. Breitenstein, J. Bauer, T. Trupke, and R. A. Bardos, *Progress in Photovoltaics: Research and Applications* 16, 325 (2008).
7. M. Kasemann, D. Grote, B. Walter, W. Kwapil, T. Trupke, Y. Augarten, R. A. Bardos, and E. Pink, *Progress in Photovoltaics: Research and Applications* 16, 297 (2008).
8. A. Helbig, T. Kirchartz, R. Schäffler, J. H. Werner, and U. Rau, *Solar Energy Materials and Solar Cells* 94, 979 (2010).
9. U. Rau, *IEEE Journal of Photovoltaics* 2, 169 (2012).
10. U. Rau, *Physical Review B* 76, 1 (2007).
11. R. S. Crandall, *Journal of Applied Physics* 53, 3350 (1982).
12. R. A. Street, *Advances in Physics* 593, 30 (1981).
13. T. C. M. Müller, B. E. Pieters, T. Kirchartz, R. Carius, and U. Rau, *physica status solidi (c)* 9, 1963 (2012).
14. B. Rech, T. Roschek, J. Mu, S. Wieder, and H. Wagner 66, 267 (2001).
15. S. Haas, G. Schöpe, C. Zahren, and H. Stiebig, *Applied Physics A* 92, 755 (2008).
16. D. L. Staebler and C. R. Wronski, *Applied Physics Letters* 31, 292 (1977).
17. A. K. O. Odziej, *Opto-electronics review* 12, 21 (2004).
18. B. E. Pieters, T. Kirchartz, T. Merdzhanova, and R. Carius, *Solar Energy Materials and Solar Cells* 94, 1851 (2010).

Mater. Res. Soc. Symp. Proc. Vol. 1536 © 2013 Materials Research Society
DOI: 10.1557/opl.2013.599

The dependence of the crystalline volume fraction on the crystallite size for hydrogenated nanocrystalline silicon based solar cells

K. J. Schmidt[1], Y. Lin[2], M. Beaudoin[3], G. Xia[2], S. K. O'Leary[1], G. Yue[4], and B. Yan[4]

[1]School of Engineering, The University of British Columbia, Kelowna, BC, Canada.
[2]Department of Materials Engineering, The University of British Columbia, Vancouver, BC, Canada.
[3]Advanced Materials and Process Engineering Laboratory, The University of British Columbia, Vancouver, BC, Canada.
[4]United Solar Ovonic LLC, Troy, MI, United States.

ABSTRACT

We have performed an analysis on three hydrogenated nanocrystalline silicon (nc-Si:H) based solar cells. In order to determine the impact that impurities play in shaping the material properties, the XRD and Raman spectra corresponding to all three samples were measured. The XRD results, which displayed a number of crystalline silicon-based peaks, were used in order to approximate the mean crystallite sizes through Scherrer's equation. Through a peak decomposition process, the Raman results were used to estimate the corresponding crystalline volume fraction. It was noted that small crystallite sizes appear to favor larger crystalline volume fractions. This dependence seems to be related to the oxygen impurity concentration level within the intrinsic nc-Si:H layers.

INTRODUCTION

Hydrogenated nanocrystalline silicon (nc-Si:H) based solar cells, as the low bandgap component cell in multi-junction thin film silicon solar cells, have attracted a considerable amount of attention in recent years owing to their improved long wavelength response and lower light induced degradation when contrasted with the case of hydrogenated amorphous silicon (a-Si:H) based solar cells [1]. Record solar cell and module efficiencies have been attained using a-Si:H/nc-Si:H/nc-Si:H triple-junction solar cells. Owing to its complexity as a material system, the material structure of nc-Si:H, and the role that impurities and the substrate play in influencing these material properties, are the subject of current intensive investigation. It has been observed that nc-Si:H based solar cells are more sensitive to impurities than a-Si:H based solar cells.

A recent study, by Yue et al. [2], examined the impurity profiles found within a series of n-i-p nc-Si:H based solar cells. The corresponding device performance was also probed, i.e., the fill-factor, the open-circuit voltage, the quantum efficiency, and the efficiency of the solar cells themselves were determined. In this analysis, we study the material properties of the nc-Si:H found within these solar cells using X-ray diffraction (XRD) and Raman spectroscopy. Correlations between the impurity concentration profiles and the material properties are sought.

EXPERIMENT

We considered three n-i-p nc-Si:H based solar cells for this analysis; in particular, sample numbers 21821, 21886, and 21916 were considered, these samples also being considered by Yue *et al.* [2]. These cells were deposited on stainless steel substrates with textured Ag/ZnO back reflectors. The phosphorous-doped a-Si:H n layer and the boron-doped nc-Si:H p layer were deposited using rf glow discharge, while the intrinsic nc-Si:H layer was deposited using VHF glow discharge with optimized hydrogen dilution and hydrogen dilution profiling. Through variations in the conditions within the deposition reactor, the impurity concentration profiles found within the nc-Si:H layers within these cells were found to be varied. In particular, while the boron impurity concentrations are, within the range of experimental error, very similar for all of the cells, the oxygen impurity concentration profiles within the intrinsic nc-Si:H layers varied between $2 - 3 \times 10^{19}$, $1 - 2 \times 10^{19}$, and less than 10^{19} cm^{-3}, for samples 21821, 21886, and 21916, respectively. Further details regarding the deposition process and the impurity levels are provided elsewhere [2]. Powder θ-2θ XRD scans are obtained using a Philips X'Pert MRD system, with a monochromatic X-ray wavelength of 1.5406 Å. The corresponding mean crystallite sizes are estimated using Scherrer's equation; we focus on the (220) peak for the purposes of this analysis. The Raman spectra are obtained from a high-resolution confocal micro-Raman system, i.e., a LabRam HR by Horiba Scientific. The excitation wavelength used is 442 nm (blue light). These Raman spectra are decomposed into their constituent peaks in order to determine the corresponding crystalline volume fraction following a process of baseline correction.

RESULTS AND DISCUSSION

In Figure 1, we plot a representative XRD spectrum corresponding to one of the samples we consider. We note that there is a distinct Si (220) peak located at about 47° [3-6]. Several other peaks, corresponding to both the film and the substrate are also identified on Figure 1: these peaks are related to the ITO, ZnO, and the stainless steel substrate. The c-Si (220) peaks, observed for all of the samples we consider, are fit using a Lorentzian curve, with a linear baseline correction. Note that the ZnO (102) peak is directly adjacent to the Si (220). Scherrer's equation was applied to the observed XRD c-Si peak widths, and the obtained mean crystallite sizes, d_{XRD}, were found to vary between 32 and 40 nm. These results are presented in Table 1. Scherrer's equation, which is derived through consideration of the constructive and destructive interference over parallel planes of atoms [7], asserts that the peaks associated with a diffraction pattern are broadened by an amount that is inversely proportional to the crystallite size. While nc-Si:H is a very complex material, with crystallites, an amorphous tissue, and grain boundaries, Scherrer's equation is often used in the analysis of this material [8, 9], despite the potential for interpretational difficulties.

In Figure 2, we plot a Raman spectrum corresponding to the representative sample considered in our XRD analysis. This spectrum was measured between 20 and 2400 cm^{-1}; in the subsequent decomposition analysis, however, we focus only on the spectrum between 420 to 540 cm^{-1} [1]. No filters were employed and low laser excitation intensity was used so as to reduce the probability of recrystallization. Distinctive peaks are observed at around 480 and 520 cm^{-1}, these corresponding to the amorphous silicon (a-Si) and crystalline silicon (c-Si) components of the Raman spectrum, respectively; these peaks are observed for all of the samples considered. There is an additional peak observed at around 510 cm^{-1}, this corresponding to the grain boundary/intermediate phase between the a-Si phase and the c-Si phase. Additional minor peaks are also observed at around 150, 310, and 380 cm^{-1}, these being representative of the vibrational modes found within a-Si [9].

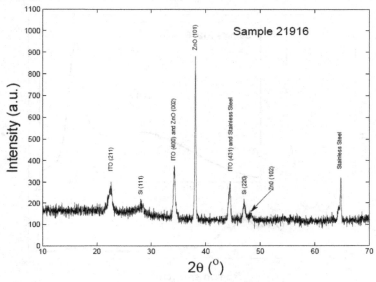

Figure 1: The XRD spectrum corresponding to a representative nc-Si:H based solar cell. The visible peaks, and their corresponding origin, are indicated in the figure.

Table 1: Summary of the mean crystallite sizes, the volume fractions, and the range of oxygen impurity levels found within the intrinsic nc-Si:H layers.

Sample	d_{XRD}	X_c	X_a	X_{GB}	Oxygen
21821	32 ± 3 nm	26 ± 1 %	74 ± 1 %	5 ± 1 %	2-3 x 10^{19} cm^{-3}
21886	35 ± 4 nm	24 ± 0.5 %	76 ± 0.5 %	4 ± 0.5 %	1-2 x 10^{19} cm^{-3}
21916	40 ± 4 nm	23 ± 1 %	77 ± 1 %	3 ± 1 %	<10^{19} cm^{-3}

In the inset to Figure 2, we focus on this representative Raman spectrum between 420 and 540 cm^{-1}. In order to determine the crystalline volume fraction of the sample, we decompose this Raman spectrum into its constituent components, assuming peaks at 480, 510, and 520 cm^{-1}; we assume symmetrical Gaussian/Lorentzian peaks, following the procedure suggested by Tay *et al.* [10]. We note that there are substantial a-Si and c-Si components to this Raman spectrum. The approach of Tsu *et al.* [11] and Bustarret *et al.* [12] is employed for the determination of the crystalline volume fraction; this approach is commonly employed in the literature. Building upon the results of Tsu et al. [11] and Bustarret et al. [12], Han et al. [13] suggest that the crystalline volume fraction, $X_c = (I_c + I_{GB})/[I_c + I_{GB} + y(L)I_a]$, where $I_c + I_{GB}$ represents the integrated comp-

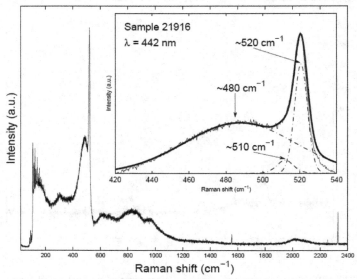

Figure 2: The Raman spectrum corresponding to a representative nc-Si:H based solar cell. The amorphous and crystalline components have been highlighted with the dashed lines at around 480 and 520 cm^{-1}, respectively. Baseline correction has not been employed. We suspect that these results will not be accurate for the lower wavenumbers, i.e., less than 150 cm^{-1}, due to input laser interference. In the inset, the representative Raman spectrum and its decomposed components at 480, 510, and 520 cm^{-1} are depicted; the excitation source is a blue laser. Baseline correction was employed for this analysis.

onents of the 510 and 520 cm^{-1} peaks, and I_a corresponds to the integrated 480 cm^{-1} peak. The ratio of the cross section for the amorphous-to-crystalline phase, $y(L)$, is set to unity for the sake of simplicity. Similar peak decompositions are performed for the other two nc-Si:H based solar cells. Table 1 summarizes the crystalline volume fractions found for the three nc-Si:H based solar cell samples, as well as the amorphous, X_a, and interface, X_{GB}, contributions.

We have analyzed the attenuation corresponding to both the X-rays generated by the XRD measurement system and the Raman source that was used and have confirmed that in both cases it is the intrinsic layer of the nc-Si:H based solar cell that is probed. The doping layer that forms the top contact is only 15-20 nm thick, while the penetration depths corresponding to the X-rays and the Raman laser were found to be 15 and 0.125 μm, respectively, the intrinsic nc-Si:H layer underneath this top contact being about 3 μm in thickness; these thicknesses are known through a comparison with the impurity concentration profile results of Yue et al. [2]. Thus, while the entire intrinsic nc-Si:H layer is being probed by the X-rays, only the top 5% of the intrinsic nc-Si:H layer is actually being probed by the Raman laser. Clearly, if there are inhomogeneities in the intrinsic nc-Si:H layers, beyond the range of the Raman laser, these would be undetectable.

In Figure 3, we plot the crystalline volume fraction as a function of the crystallite size. It is noted that there appears to be an increase in the crystalline volume fraction as the mean crystallite sizes are diminished. It is also observed that the crystalline volume fraction diminishes along with the oxygen impurity content while the crystalline sizes increase. We speculate that the decrease in the crystalline volume fraction with increased mean crystallite size may be related to the ability of smaller crystallites to be arranged with a greater packing density than their larger counterparts. A similar result was suggested by Funde et al. [14], although it should be noted that very different ranges of the crystalline volume fraction and the crystallite sizes were considered in their analysis. The reason for the correlation with the oxygen level in our present analysis remains unknown.

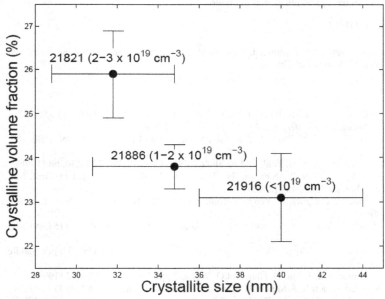

Figure 3: The crystalline volume fraction as a function of the mean crystallite size for the three nc-Si:H based solar cells considered in this analysis. The corresponding oxygen impurity levels found within the intrinsic nc-Si:H layers are indicated in the brackets. Error bars associated with these measurements are indicated.

CONCLUSIONS

We have performed an analysis on three nc-Si:H-based solar cells. We measured the XRD and Raman spectra corresponding to all three samples. The XRD results, which display a number of c-Si-based peaks, were used to measure the mean crystallite sizes. Through a peak decomposition procedure, the Raman results are used to estimate the corresponding crystalline volume fraction. It is noted that smaller crystallite sizes appear to favor larger crystalline volume fractions. Further analysis, however, will be required in order to confirm this observation.

There are several difficulties with the experimental results as presented, thus far. First, the changes observed in the crystalline volume fraction are relatively modest. Even though the error bars themselves are small compared with the actual variations in the data, this does raise some concern. Second, there are interpretational difficulties associated with the determination of the crystalline volume fraction from the Raman spectra, i.e., the decomposition procedure employed and how deep the Raman laser penetrates into the intrinsic nc-Si:H layer. A full error analysis, further samples, a higher Raman wavelength, i.e., a more deeply penetrating Raman laser source, and further scrutiny of the experimental procedure would need to be considered in order to confirm this observation. A connection with the device performance would also be interesting. This will have to be pursued in the future.

ACKNOWLEDGMENTS

The authors gratefully acknowledge financial support from the Natural Sciences and Engineering Research Council of Canada.

REFERENCES

1. B. Yan, G. Yue, L. Sivec, C.-S. Jiang, Y. Yan, K. Alberi, J. Yang, and S. Guha, in *35th IEEE Photovoltaic Specialists Conference, PVSC 2010*, 3755 (2010).
2. G. Yue, B. Yan, L. Sivec, Y. Zhou, J. Yang, and S. Guha, Sol. Energy Mater. Sol. Cells **104**, 109 (2012).
3. U. Kroll, J. Meier, P. Torres, J. Pohl, and A. Shah, J. Non-Cryst. Solids **227-230**, 68 (1998).
4. A. V. Shah, J. Meier, E. Vallat-Sauvain, N. Wyrsch, U. Kroll, C. Droz, and U. Graf, Sol. Energy Mater. Sol. Cells **78**, 469 (2003).
5. G. Yue, B. Yan, G. Ganguly, J. Yang, S. Guha, and C. W. Teplin, Appl. Phys. Lett. **88**, 263507 (2006).
6. G. Yue, L. Sivec, B. Yan, J. Yang, and S. Guha, Mater. Res. Soc. Symp. Proc. **1153**, A10-05 (2009).
7. B. D. Cullity and S. R. Stock, *Elements of x-ray diffraction* (Prentice-Hall, Upper Saddle River, 2001).
8. G. Yue, J. D. Lorentzen, J. Lin, D. Han, and Q. Wang, Appl. Phys. Lett. **75**, 492 (1999).
9. E. Vallat-Sauvain, U. Kroll, J. Meier, A. Shah, and J. Pohl, J. Appl. Phys. **87**, 3137 (2000).
10. L.-L. Tay, D. J. Lockwood, J.-M. Baribeau, M. Noël, J. C. Zwinkels, F. Orapunt, and S. K. O'Leary, Appl. Phys. Lett. **88**, 121920 (2006).
11. R. Tsu, J. Gonzalez-Hernandez, S. S. Chao, S. C. Lee, and K. Tanaka, Appl. Phys. Lett. **40**, 534 (1982).
12. E. Bustarret, M. A. Hachicha, and M. Brunel, Appl. Phys. Lett. **52**, 1675 (1988).
13. D. Han, J. D. Lorentzen, J. Weinberg-Wolf, L. E. McNeil, and Q. Wang, J. Appl. Phys. **94**, 2930 (2003).
14. A. M. Funde, N. A. Bakr, D. K. Kamble, R. R. Hawaldar, D. P. Amalnerkar, and S. R. Jadkar, Sol. Energy Mater. Sol. Cells **92**, 1217 (2008).

Mater. Res. Soc. Symp. Proc. Vol. 1536 © 2013 Materials Research Society
DOI: 10.1557/opl.2013.600

Carrier Lifetime Measurements by Photoconductance at Low Temperature on Passivated Crystalline Silicon Wafers

Guillaume Courtois[1,2], Bastien Bruneau[2], Igor P. Sobkowicz[1,2], Antoine Salomon[1] and Pere Roca i Cabarrocas[2]

[1]Total New Energies, R&D Division, Tour Michelet, 24 Cours Michelet – La Défense 10
92069 Paris La Défense Cedex, France
[2]LPICM, CNRS – Ecole Polytechnique, Bât 406, Route de Saclay
91128 Palaiseau Cedex, France

ABSTRACT

We propose an implementation of the PCD technique to minority carrier effective lifetime assessment in crystalline silicon at 77K. We focus here on (n)-type, FZ, polished wafers passivated by a-Si:H deposited by PECVD at 200°C. The samples were immersed into liquid N_2 contained in a beaker placed on a Sinton lifetime tester. Prior to be converted into lifetimes, data were corrected for the height shift induced by the beaker. One issue lied in obtaining the sum of carrier mobilities at 77K. From dark conductance measurements performed on the lifetime tester, we extracted an electron mobility of 1.1×10^4 cm². V^{-1}.s^{-1} at 77K, the doping density being independently calculated in order to account for the freezing effect of dopants. This way, we could obtain lifetime curves with respect to the carrier density. Effective lifetimes obtained at 77K proved to be significantly lower than at RT and not to depend upon the doping of the a-Si:H layers. We were also able to experimentally verify the expected rise in the implied V_{oc}, which, on symmetrically passivated wafers, went up from 0.72V at RT to 1.04V at 77K under 1 sun equivalent illumination.

INTRODUCTION

Properties of passivated crystalline wafers involved in solar cell manufacturing have been widely investigated at room temperature (RT). For instance, photoconductance has been widely used since the mid-1990s in order to assess minority carriers effective lifetimes. Among all possible candidates, hydrogenated amorphous silicon (a-Si:H) has proved to ensure very effective surface passivation, hence very long carrier lifetimes obtained in such passivated wafers and high open-circuit voltages in silicon heterojunction solar cells [1]. Yet, passivation mechanisms into play are not yet completely understood, and low temperature measurements may contribute to sharpen knowledge on that topic. Whilst a slight increase in bandgap alongside with a sharp increase in implied V_{oc} at low temperature are well predicted by semiconductor physics, alterations in carrier effective lifetimes are difficult to account for. Experimental studies carried out so far on comparable materials, relying either on microwave-detected photoconductance decay (PCD) [2], or modulated photoluminescence (PL) [3], [4] agree on a decrease in effective lifetime as the temperature diminishes down to 70K. We report here on effective lifetime measurements performed by inductively coupled PCD at 77K on (n)-type mono-crystalline wafers passivated by a-Si:H.

EXPERIMENTAL DETAILS

(n)-type, FZ, 280μm-thick, double-side polished wafers were used in our experiments. Once the wafers had undergone a HF dip, a-Si:H layers were deposited on both sides by RF-PECVD at 200°C. For ensuring surface passivation, a thin intrinsic layer was always inserted between the wafer and the doped a-Si:H layers. Starting from identical wafers whose nominal resistivity was of 3Ω.cm, different structures were compared: symmetrical stacks featuring either (n+) or (p)-doped a-Si:H layers, known to lead to distinct surface passivation [5], as well as the actual silicon heterojunction solar cell precursor structure composed of (p) front layer and highly-doped (n+) back surface field (see Figure 1).

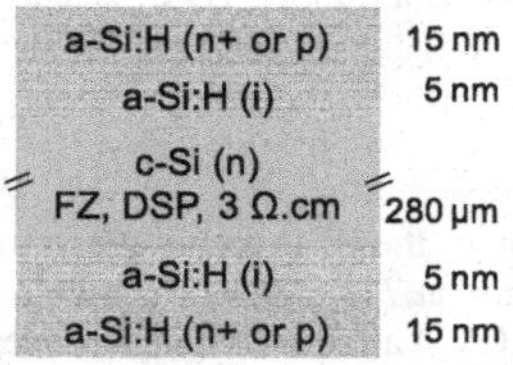

a-Si:H (n+ or p)	15 nm
a-Si:H (i)	5 nm
c-Si (n) FZ, DSP, 3 Ω.cm	280 μm
a-Si:H (i)	5 nm
a-Si:H (n+ or p)	15 nm

Figure 1. Structure of the investigated samples.

Measurements were performed by inductively coupled PCD, in which the excess photoconductance exhibited by the wafer when excited by a flash lamp is assessed by an inductive coil embedded in the sample holder. We implemented this technique to low temperature measurements by immersing the passivated wafers into liquid nitrogen contained into a beaker placed on the stage of a WCT-120 lifetime tester by Sinton Instruments. In order for the measuring coil to keep warm enough, a thin insulating material had to be inserted below the beaker. To enhance the precision, we always performed an averaging on several flashes. The samples were excited by short flashes and the data analysis was operated in quasi-transient regime, which amounts to assume that excess carriers have been photogenerated before the data acquisition starts. The effective lifetime τ_{eff} is therefore given at every instant by the following expression (provided lifetimes turn out greater than, typically, 200μs):

$$\frac{\Delta p}{\tau_{eff}} = -\frac{\partial \Delta p}{\partial t} \tag{1}$$

where Δp stands for the excess minority carrier density – say, the density of photogenerated holes. It is assumed equal to the excess majority carrier density Δn. In practice, from the discrete experimental values, for a given injection level Δp, τ_{eff} is computed through the local slope of the Δp versus time curve. Note that Δp is beforehand derived from the measured excess photoconductance ΔS_0 according to:

$$\Delta S_0^T = q * \Delta p^T * (\mu_e^T + \mu_h^T) * W \tag{2}$$

where μ_e and μ_h stand for the electron and hole mobilities, respectively, and W for the thickness of the wafer, whilst the exponent T indicates the temperature-dependent quantities.

Explicit methodology

Since our wafers were not placed directly upon the lifetime tester stage but in a beaker, raw excess photoconductance data had first to be corrected for the relevant height shift prior to be converted into injection levels Δp. For that purpose, we implemented a method detailed in [6] by fitting the photoconductance drop induced by the height shift (h) with an exponential decay:

$$\Delta S^{T}(h) = \Delta S_{0}^{T} * \exp(-\alpha^{T} h) \tag{3}$$

in which h shall not exceed a few millimeters for the expression to stay valid.
We checked that the decay coefficient α was independent from the investigated wafer, as stated in [5]. Once this coefficient determined, the actual excess photoconductance ΔS_{0}^{T} values were retrieved at every instant from the ΔS^{T} values.

Furthermore, not only is the response to light excitation altered when the temperature changes, but intrinsic properties under equilibrium are deeply altered as well. On the first hand, the doping density (N_D) had to be recalculated in order to account for the freezing effect of dopants. For instance, a 3Ω.cm bulk resistivity corresponds to a doping density of 1.6×10^{15} cm^{-3} at 300K, which shrinks to 1.0×10^{15} cm^{-3} at 77K [7].
On the other hand, one issue lied in obtaining the carrier mobilities at 77K for the specific doping of the wafers into question. We addressed this issue by further extrapolating the conductance values provided by the lifetime tester. Indeed, before each lifetime measurement, a conductance assessment is automatically performed in the dark – i.e. before the flash shines. Like exposed above, raw dark conductance values obtained with the wafer into the beaker had first to be corrected for the height shift according to Equation 4 prior to extracting the electron mobility from Equation 5. This procedure was validated at RT.

$$S^{T}(h) = S_{0}^{T} * \exp(-\beta^{T} h) \tag{4}$$

$$S_{0}^{T} \approx q * N_{D}^{T} * \mu_{e}^{T} * W \tag{5}$$

The electron mobility μ_e we obtained this way was equal to 1.1×10^{4} cm^2.V^{-1}.s^{-1} at 77K – showing good agreement with the calculations reported in [8] – to be compared to 1.35×10^{3} at 300K. As to the hole mobility μ_h, it was assumed to be of the order of one quarter of μ_e – just like the ratio at 300K. In our computations, the mobility sum was set constant on the entire range of investigated injection levels – contrariwise to what happens at RT, where, at high injection levels, a slight decrease from the initial value of 1.7×10^{3} cm^2.V^{-1}.s^{-1} is reckoned with.

RESULTS

Effective lifetimes

As generally observed, effective lifetimes were significantly higher at RT with (n+)-doped a-Si:H layers than with the (p) ones, owing to a likely defect formation in the intrinsic thin layer induced by the (p)-type overlayer [5]. At 77K, whatever the samples, we observed a significant drop in effective lifetime values. More precisely, for wafers symmetrically passivated by (n+)-doped a-Si:H, for which effective lifetimes up to 5.2ms at 10^{15} cm^{-3} were reached at RT, the lifetime fell to 0.92ms, whilst for their counterparts symmetrically passivated by (p)-doped a-Si:H, it sank from 1.8ms at RT to 0.97ms at 77K. The mismatch observed at 77K between the two kinds of samples is considered not significant (see Figure 2). We subsequently checked that the initial lifetimes were recovered at RT. As to cell precursors, they behaved similarly to wafers symmetrically passivated by (p)-doped a-Si:H, denoting that the carrier lifetime in an actual cell is ruled by the (p)-doped layer.

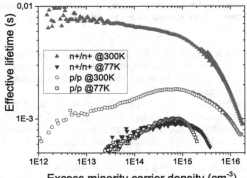

Figure 2. Lifetime measurements at 300K (red) and at 77K (blue) on wafers symmetrically coated by (i)/(n+) stacks (full symbols) and (i)/(p) a-Si:H stacks (open symbols). The similarities in the curves obtained at 77K suggest that the effective lifetime becomes bulk-dominated at this temperature.

$$\frac{1}{\tau_{eff}} = \frac{1}{\tau_{bulk}} + \frac{1}{\tau_{surface}} = \left(\frac{1}{\tau_{rad}} + \frac{1}{\tau_{Auger}} + \frac{1}{\tau_{SRH}}\right) + \frac{1}{\tau_{surface}} \qquad (6)$$

Effective lifetimes convey combination of both bulk and surface recombination phenomena, the latter overwhelming the former at RT in the high quality wafers we used. When decreasing temperature, an increase in the radiative coefficient (hence a decrease of radiative lifetime) has been established down to 77K ([9], [10]). Besides, the parameterization of Auger coefficients over the 70-400K range proposed in [11] predicts a decrease in Auger lifetime as the temperature decreases. Coming to the SRH contribution, since the probability of thermal capture decreases when decreasing temperature, the SRH lifetime is prone to increase [12], and so is the surface lifetime, since surface recombination occurs via defects (namely dangling bonds), to which the SRH description can be extended. Whatever the conclusions regarding the SRH mechanism, the fact that almost identical curves were obtained at 77K for samples initially exhibiting different surface passivation features at RT suggests that, on the investigated structures, the effective lifetime is led by bulk recombination at 77K.

Implied Voc

In addition to carrier effective lifetimes, another paramount information provided by the lifetime tester is the implied open-circuit voltage, or, in other words, the splitting of the quasi-Fermi levels $E_{Fn} - E_{Fp}$. An upper bound for the implied V_{oc} is therefore set by the silicon bandgap value, which slightly increases from 1.12 eV at 300K to almost 1.17eV at 77 K. In the case of a (n)-type wafer, the implied V_{oc} reads as follows:

$$V_{oc} = \frac{E_{Fn} - E_{Fp}}{q} = \frac{kT}{q}\ln\left(\frac{(N_D + \Delta p)\Delta p}{n_i^2}\right) \qquad (7)$$

The intrinsic carrier density n_i obviously depends strongly upon the temperature and drastically falls from about 8.6×10^9 cm^{-3} at 300K to about 10^{-20} cm^{-3} at 77K [6]. Fermi-Dirac statistics for electrons gives:

$$E_c - E_{Fn} = kT\ln\left(\frac{N_C}{N_D + \Delta p} - 1\right) \qquad (8)$$

where N_C is the density of states in the conduction band. The temperature dependence of the logarithm argument being dominated by N_C, which decreases [13], the quasi-Fermi level for electrons tends towards the conduction band level as kT decreases. The symmetrical argument applied to holes indicates that $E_{Fn} - E_{Fp}$ will tend towards the band gap at low temperatures. A comparison between implied V_{oc} at 300K and 77K is displayed on Figure 3. On wafers symmetrically passivated by (n+)- a-Si:H, the implied V_{oc} rose from 0.72V at RT to 1.04V at 77K under 1 sun equivalent, which showed good agreement with the results reported in [4] on (p)-type mono-crystalline wafers also passivated by a-Si:H.

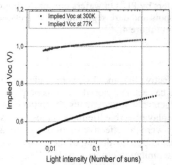

Figure 3. Implied V_{oc} obtained at 300K and 77K on a wafer symmetrically coated by (i)/(n+) a-Si:H stacks. Under 1 sun equivalent, it rose from 0.72 to 1.04V.

Wafer alteration

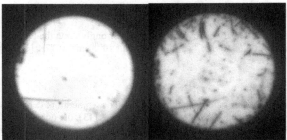

Figure 4. PL images of a 4-inches wafer symmetrically coated by (i)/(n+) a-Si:H stacks, before any dip in liquid nitrogen (left) and after numerous dips in liquid nitrogen (right). Both images were taken at room temperature under an 810nm laser excitation.

Interestingly, a side effect induced by the thermal shocks undergone by our samples was revealed when monitoring these wafers by PL. On wafers purposely dipped many times in liquid N_2, numerous initially absent dark spots and dark segments, i.e. defects, were visible on PL images taken afterwards, as shown in Figure 4.

CONCLUSION

A protocol for measuring effective lifetimes by photoconductance decay at 77K on passivated crystalline silicon wafers has been proposed. For this purpose, we extended a work on the height shift effect on photoconductance measurements. Once the majority carrier mobility assessed at 77K, we were able to display effective lifetimes versus excess minority carrier density at this temperature. Not only were the effective lifetimes significantly shorter at 77K than at RT, but they turned out independent upon the doping of the passivation layers, suggesting a bulk limitation instead of the surface limitation observed at RT. Finally, accordingly to theoretical predictions, a rise in implied open-circuit voltage was highlighted at low temperature.

ACKNOWLEDGMENTS

The authors would like to thank Joaquim Nassar for fruitful discussions and Julien Penaud for help in acquiring the PL images.

REFERENCES

1. S. De Wolf, A. Descoeudres, Z. C. Holman and C. Ballif, *Green* **2**,7 (2012).
2. R. Leadon and J. A. Naber, *Journal of Applied Physics* **40**, 2633 (1969).
3. T. Trupke, J. Zhao, A. Wang, R. Corkish and M. A. Green, *Applied Physics Letters* **82**, 2996 (2003).
4. S. Tardon, PhD. Thesis, Universität Oldenburg (2006).
5. S. De Wolf and M. Kondo, *Journal of Applied Physics* **105**, 103707 (2009).

6. W. Favre, L. Bettaieb, J. Després, J. Alvarez, J.-P. Kleider, Y. Le Bihan, Z. Djebbour and D. Mencaraglia, *Proceedings of the 26th EUPVSEC* (2011).

7. B. Van Zeghbroeck, "Principles of Semiconductor Devices, 2.6.4.4", Online http://ecee.colorado.edu/~bart/book/welcome.htm (2011). [Accessed 19 March 2013].

8. S. S. Li and W. R. Thurder, *Solid-State Electronic* **20**, 609 (1977).

9. H. Schlangenotto, H. Maeder and W. Gerlach, *Phys. Status Solidi A* **21**, 357 (1974).

10. T. Trupke, M. A. Green, P. Würfel, P. P. Altermatt, A. Wang, J. Zhao and R. Corkish, *Journal of Applied Physics* **94**, 4930 (2003).

11. P. P. Altermatt, J. Schmidt, G. Heiser and A. G. Aberle, *Journal of Applied Physics* **82**, 4938 (1997).

12. A. Schenk, *Solid-State Electronics* **35**, 1585 (1992).

13. S. M. Sze and K. K. Ng, *Physics of Semiconductor Devices*, 3rd ed. (Wiley Intersciences, 2007) p18.

Mater. Res. Soc. Symp. Proc. Vol. 1536 © 2013 Materials Research Society
DOI: 10.1557/opl.2013.752

Characterization of Boron Doped Amorphous Silicon Films by Multiple Internal Reflection Infrared Spectroscopy

N. Ross[1], K. Shrestha[2], O. Chyan[1], C. L Littler[2], V. C. Lopes[2], and A. J. Syllaios[2]

[1]Department of Chemistry and [2]Department of Physics, University of North Texas, Denton, TX 75203, USA

ABSTRACT

In this study, we employed Multiple Internal Reflection Infrared Spectroscopy (MIR-IR) to characterize chemical bonding structures of boron doped hydrogenated amorphous silicon (a-Si:H(B)). This technique has been shown to provide over a hundred fold increase of detection sensitivity when compared with conventional FTIR. Our MIR-IR analyses reveal an interesting counter-balance relationship between boron-doping and hydrogen-dilution growth parameters in PECVD-grown a-Si:H. Specifically, an increase in the hydrogen dilution ratio (H_2/SiH_4) was found to cause the increase in the Si-H bonding and a decrease in the B-H and SiH_2 bonding, as evidenced by the changes in corresponding IR absorption peaks. In addition, although a higher boron dopant gas concentration was seen to increase the BH and SiH_2 bonding, it also resulted in the decrease of the most stable SiH bonding configuration. The new chemical bonding information of a-Si:H thin film was correlated with the various boron doping mechanisms proposed by theoretical calculations.

INTRODUCTION

Thin amorphous silicon and its alloy films are widely used in important commercial applications including thin-film transistor and photovoltaic solar cell. For uncooled microbolometer infrared detectors, optimizing the key properties of film morphology, temperature coefficient of resistance and electrical conductivity is crucial for the detector performance. Extensive characterization has been carried out on the effects of PECVD growth conditions on optical and electrical properties of amorphous silicon [1-4]. However, the chemical bonding structure is less studied. In this work, we have used MIR-IR spectroscopic tool to investigate the chemical bonding of hydrogenated amorphous silicon (a-Si:H) thin film and explore its boron doping mechanism.

EXPERIMENT

Samples of a-Si:H thin films were grown in a PECVD system using capacitively coupled 13.56 MHz plasmas with the substrate on the ground electrode. Source gases for p-type a-Si:H growth were silane (SiH_4) and hydrogen in argon, and boron trichloride (BCl_3). MIR-IR was used to characterize the chemical bonding structure differences due to variation of H dilution and B doping during PECVD deposition. MIR-IR utilizes silicon wafer itself as an IR waveguide for attenuated total reflection (ATR) to achieve sub monolayer detection sensitivity [5]. Favorable difference of refractive indices of Si wafer and its air interface enables multiple total internal reflections (ca. 80 reflections from a 60 mm Si ATR crystal), which greatly enhances IR measuring sensitivity. The a-Si:H (ca. 50 nm) coated silicon wafers were fabricated on to an ATR optical element (60 x 10 x 0.7 mm, 45° bevel angle) by mechanical polishing [5]. For Si<100> background ATR coupons, Standard Cleaning (SC1) solution was used to remove any organic contamination from the surface followed by etching in 0.5% HF solution. IR spectra were recorded on a Nicolet IS 50 FTIR spectrometer under constantly purged with dry air ($CO_2 <$ 1 ppm). Both transmission (TIR) and MIR-IR infrared spectra were measured. All spectra were collected at 2 cm^{-1} resolution and are the average of 100 individual spectra. The detection limit of MIR-IR is ca. 0.2 mabs with an error margin of < 3%.

RESULTS

The infrared absorption spectra (MIR-IR and TIR) of an a-Si:H thin film sample grown over Si_xN_y:H coated Si<100> are shown in Fig. 1. MIR-IR demonstrates over a hundred fold increase of detection sensitivity compared with conventional TIR. The multiple sampling enabled by more than 80 total internal reflections greatly enhances the measuring sensitivity of MIR-IR technique. Previously, we utilized MIR-IR to monitor hydrogen passivation on Si<100> surface with sub-monolayer measuring sensitivity [6]. The MIR-IR spectrum, Fig. 1, reveals well-resolved IR absorption peaks associated with Si-H (1989 cm^{-1}) and B-H (2465 cm^{-1}) bonding modes originated form a-Si:H thin film [7]. The observed infrared absorption peaks of N-H (3340 cm^{-1}) and SiH_x (2000 cm^{-1}) bonding modes can be assigned to the underlying Si_xN_y:H layer on Si<100> substrate.

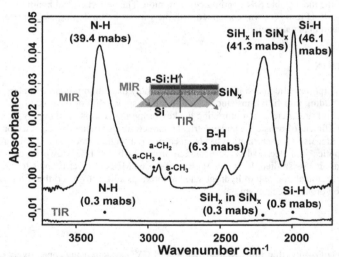

Fig. 1 MIR-IR (top) and TIR (bottom) spectra of a 50 nm a-Si:H thin film deposited on Si_xN_y:H.

Effects of boron doping ratio and hydrogen dilution growth parameters on PECVD a-Si:H thin film were studied by MIR-IR. Fig. 2a & 3a show that increasing of the boron doping ratio ([BCl_3]/[SiH_4]) results in a continued, but a more pronounced decrease of SiH bonds. In figure 2b and 3b, when the hydrogen dilution ratio ([H_2]/[SiH_4]) is increased in the boron doped a-Si:H deposition process, there is an increase in the Si-H bond formation and a decrease in the B-H bond formation. Also, the N-H (3340 cm^{-1}) and SiH$_x$ (2000 cm^{-1}) infrared absorption peaks from underlying substrate Si$_x$N$_y$:H remain constant as expected in figure 2a and 2b.

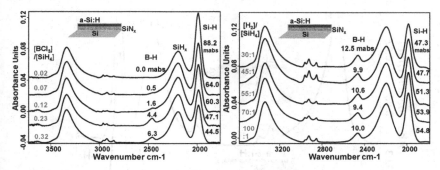

Fig. 2   MIR-IR spectra of 50 nm a-Si:H thin films deposited with different PECVD grown parameters a) boron doping [BCl_3]/[SiH_4] ratio, b) hydrogen dilution [H_2]/[SiH_4] ratio.

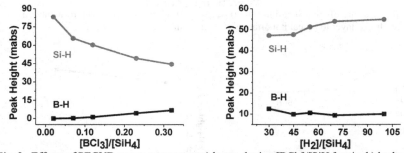

Fig. 3   Effects of PECVD grown parameters a) boron doping [BCl_3]/[SiH_4] ratio. b) hydrogen dilution [H_2]/[SiH_4] ratio on IR absorption peaks heights of B-H and Si-H in a-Si:H thin films.

As shown in Fig. 1, the underlying Si$_x$N$_y$:H layer exhibits a strong SiH$_x$ absorption peak centered at 2000 cm^{-1} that can mask the important chemical bonding information of PECVD a-Si:H film. To examine more closely the IR absorption at the region of 1800-2400 cm^{-1}, a-Si:H thin films were deposited directly on Si<100> substrate. As Fig. 4 shows, by removing the Si$_x$N$_y$:H layer, two additional IR peaks at 2105 cm^{-1} and 2260 cm^{-1} were observed from a-Si:H thin film and can be assigned to SiH$_2$ and O-SiH$_x$ bonding modes respectively [8]. Effects of doping ratio and hydrogen dilution on a-Si:H thin film deposited on Si<100> substrate were also studied by MIR-IR. Fig. 5a & 5b confirm that an interesting counter-balance relationship exists between boron-doping and hydrogen-dilution growth parameters in PECVD-grown a-Si:H. Specifically, an increase in the hydrogen dilution ratio causes the increase in the Si-H bonding and a decrease in the B-H and SiH_2 bonding. In comparison, an increase in B doping ratio results in the increase of B-H bonding and decrease of Si-H bonding mode.

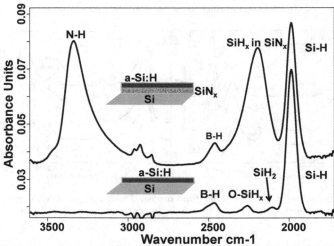

Fig. 4   MIR-IR spectra of an a-Si:H thin film a) with a Si_xN_y:H underlying layer b) without a Si_xN_y:H underlying layer to reveal important SiH_2 and O-SiH_x peaks of a-Si:H .

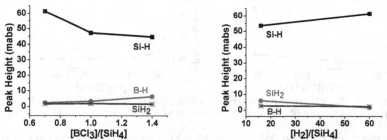

Fig. 5   Effects of PECVD grown parameters: $[BCl_3]/[SiH_4]$ ratio and $[H_2]/[SiH_4]$ ratio on IR absorption peaks heights of B-H, SiH_2 and Si-H in a-Si:H thin films deposited on Si<100> .

DISCUSSION

Deposition Condition Effects on Chemical Bonding Structure

MIR-IR reveals four IR-active bonding modes including B-H, O-SiHx, SiH_2 and Si-H in the 50 nm a-Si:H thin film grown by PECVD. An interesting counter-balance relationship between boron-doping and hydrogen-dilution growth parameters was consistently observed in all a-Si:H films deposited on Si_xN_y:H or Si<100>. A simplistic explanation can be described as there is a competition between B and Si toward H during PECVD growth process. Based on bond energy calculations, B-H (389 KJ/mole) is more energetically favorable than Si-H (318 KJ/mole) [9]. Our MIR-IR data, Fig 3a vs. 3b, show that an increase of boron-doping ratio gives a more pronounced decrease of Si-H bonds and steady increase of B-H bonds. However, PECVD growth process involves energetic ions, reactive radicals and neutral moieties from precursors that interacting with constant evolving surface with various active sites. More work is needed to give a reliable account of the observed counter-balance relationship. On the other hand, optimization of B-doping in a-Si:H films to achieve proper balance of temperature coefficient of resistance and conductivity is critical important for the infrared microbolometer imaging application. Our MIR-IR data bring new insights toward better understanding of detailed B-doping mechanism.

It is well known that B-doping in amorphous silicon (a-Si) is low (ca. 10%) as compared to crystalline silicon (c-Si) [4]. The repeating crystal structure in c-Si favors B atoms to assume the fourfold configuration as its surrounding c-Si lattice. The resulting substitutional B doping provide mobile hole carriers to shift the Fermi-level and achieve active doping. However, relaxation effects in the amorphous matrix can significantly reduce the dominance of the fourfold substitutional doping. Several hypotheses have been proposed for explaining the cause of low doping efficiency of B in a-Si and a-Si:H. Most of B could be incorporated into threefold coordinate sites that are inert and non-doping [4]. In addition, localized midgap states commonly associated with dangling bonds as well as tail states in the valence and conduction bands could contribute to the non-active B doping configuration [10]. Recently, Santos et al. used *ab initio* simulations to propose the presence of intrinsic hole traps associated with highly distorted angles in the amorphous matrix [10]. Consequently, even the substitutional B doping in fourfold configuration can lose its hole carriers to these hole traps nearby. Hydrogen as the key compositional ingredient of a-Si:H and has been associated with the inactive B doping. Boyce and Ready [11] used NMR measurements to show one-half of all B atoms has a neighboring H atoms about 1.4 Å away and suggested H-passivation on B as the likely cause of non-active B doping. Fedders and Drabold proposed based on simulation that B(3,1) (a B with three Si and one H neighbors) is the most favorable doping configuration energetically [12]. They also proposed that B-H pair formation, similar to c-Si, could disable the B-doping in a-Si:H by inserting a H atom between a B atom and one of its Si nearest neighbors.

Our MIR-IR analyses reveal detailed chemical bonding characteristics of B-doped a-Si:H thin film to collaborate with various theoretical predictions. As shown in Fig. 5, B-doping generates mainly B-H bonding in the a-Si:H matrix observed as a vibrational mode at 2465 cm^{-1}. No higher H substitution on B, like BH_2 and BH_3 vibrational modes at 2500-2565 cm^{-1}[13], were observed. In addition, no three-centered-two-electron Si-H-B bridge formation was observed at 1850 cm^{-1} in the PECVD a-Si:H thin films up to B doping ratio of 0.32. Therefore, H-passivation via the B-H pair formation is not the main cause of inefficient B doping in our a-Si:H thin film. Previously, the resistivity a-Si:H thin film were shown to reach saturation after B doping ratio of 0.13 and further increase in B dopant concentration do not yield lower resistivity [2]. However, MIR-IR spectroscopic anslyses show B-H bonding in a-Si:H continues to increase up to B doping ratio of 0.32, Fig. 2a. The result suggests that the part of the inactive B-doping coordination configuration should include B-H bonding. We propose that the threefold coordination of B (2,1) (a B with two Si and one H neighbors) contributes to the inactive B-doping in PECVD grown a-Si:H thin films based on MIR-IR characterization.

SUMMARY

MIR-IR characterization of PECVD grown a-Si:H thin film shows well-resolved bonding modes of B-H, O-SiHx, SiH_2 and Si-H. An interesting counter-balance relationship between boron-doping and hydrogen-dilution growth parameters was observed. MIR-IR data suggest a competition of B and Si to capture H during the PECVD growth process. Threefold coordination of B (2,1) (a B with two Si and one H neighbors) could contribute to the inactive B-doping in a-Si:H thin film.

ACKNOWLEDGMENTS
This work was done under the ARO grant W911NF-10-1-0410, William W. Clark Program Manager.

REFERENCES
1. A.J. Syllaios, S.K. Ajmera, G.S. Tyber, C. Littler, R.E. Hollingsworth. Mater. Res. Soc. Symp. 1153 (2009)
2. S.K. Ajmera, A.J. Syllaios, G.S. Tyber, M.F. Taylor, R.E. Hollingsworth. Proc. of SPIE. 7660 (2010)
3. D.B. Saint John, H.B. Shin, M.Y. Lee, S.K. Ajmera. A.J. Syllaios, E.C. Dickey, T.N. Jackson, N.J. Podraza. J. Appl. Phys. 110 (2011)
4. R.A. Street, *Hydrogenated Amorphous Silicon*. (Cambridge University Press, UK, 2002)
5. D. Bhattacharyya, K. Pillai, O. Chyan, L. Tang, R.B. Timmons, Chem. Mater. 19, 2222 (2007)
6. O. Chyan, J. Wu, J.J. Chen. Appl. Spec. 51, 1905 (1997)
7. C. Guanghua, Z. Chenzhi, Z. Angqing, C. Jinlong and C. Wei. Phys. Stat. Sol. 96, K187 (1986)
8. C. Liu, S. Palsule, Yi, S. Gangopadhyay. Phys. Rev. B 49, (1994)
9. T.L. Cottrell, *The Strengths of Chemical Bonds*, 2nd ed. (Butterworths Scientific Publication, London, 1958)
10. Santos, P. Castrillo, W. Windl, D.A. Drabold, L. Pelaz and L.A. Marques, Phys. Rev. B 81, (2010)
11. J. B. Boyce and S. E. Ready, Phys. Rev. B 38, 11008 (1998)
12. P.A. Fedders and D.A. Drabold, Phys. Rev. B 54, 1864 (1997)
13. G. Socrates. Infrared and Raman Characteristic Group Frequencies. (John Wiley & Sons, USA, 2001, 247-253)

Mater. Res. Soc. Symp. Proc. Vol. 1536 © 2013 Materials Research Society
DOI: 10.1557/opl.2013.919

Emission spectra study of plasma enhanced chemical vapor deposition of intrinsic, n^+, and p^+ amorphous silicon thin films

I-Syuan Lee and Yue Kuo
Thin Film Nano & Microelectronics Research Laboratory, Texas A&M University,
College Station, TX, 77843-3122

ABSTRACT

The PECVD intrinsic, n^+, and p^+ a-Si:H thin film deposition processes have been studied by the optical emission spectroscope to monitor the plasma phase chemistry. Process parameters, such as the plasma power, pressure, and gas flow rate, were correlated to SiH^*, H_α^*, and H_β^* optical intensities. For all films, the deposition rate increases with the increase of the SiH^* intensity. For the doped films, the H_α^*/SiH^* ratio is a critical factor affecting the resistivity. The existence of PH_3 or B_2H_6 in the feed stream enhances the deposition rate. Changes of the free radicals intensities can be used to explain variation of film characteristics under different deposition conditions.

INTRODUCTION

Plasma enhanced chemically vapor deposited (PECVD) a-Si:H thin films have been widely used in thin film transistors (TFTs), p-i-n diodes, solar cells, optoelectronic devices, etc. [1,2]. Process parameters, such as the power, pressure, temperature, and feed gas stream, affect the deposition rate and electrical properties of the film [3,4]. The optical emission spectroscopy (OES) is a powerful diagnosis tool to characterize the PECVD process [5]. For example, for the a-Si:H deposition using the SiH_4/H_2 feed stream, excited species such as SiH^* (414 nm), H_α^* (656 nm), and H_β^* (486 nm) are important components in the plasma phase. It was reported that the H_α^*/SiH^* intensity ratio could be related to the crystallinity in the film [6]. In addition, the H_β^*/H_α^* ratio could be used to estimate the electron temperature (T_e) [7]. The high T_e favors the microcrystalline silicon (μc-Si:H) formation. However, when the T_e is too high, it leads to instability of the film's microstructure [8]. In this paper, the OES is used to monitor the plasma phase chemistry during PECVD intrinsic (i-), n^+, and p^+ thin film depositions. Changes of the SiH^*, H_α^*, and H_β^* free radicals intensities with respect to the plasma power, pressure, and feed gas flow rate are investigated. For the doped film, the relationship between the free radical intensity and the resistivity is studied.

EXPERIMENTAL

The i-, n^+, and p^+ a-Si:H thin films were deposited with a PECVD system that had a parallel-electrode design at 250°C on a pre-cleaned Corning 1737 glass substrate. The top electrode was driven by 13.56 MHz RF power generator with an automatic matching box. The pressure range was 150-800 mT and the power range was 80-500 W. For the i-layer deposition, SiH_4 was used as the feed gas. For the n^+ film deposition, the feed stream was composed of SiH_4 (10-60 sccm), H_2, (1000 sccm), and PH_3 (6.88% in H_2) (1-20 sccm). For the p^+ film deposition, the feed stream was composed of SiH_4 (10-35 sccm), H_2 (400-

1000 sccm), and B_2H_6 (2% in H_2) (10-100 sccm). The film's resistivity was determined from a 4-point probe (Alessi Industry). The emission spectrum in the 250-800 nm range was monitored using an OES. Before the plasma is generated, all the intensities are set to almost 0 by deducting the background. After the plasma is generated, all the intensities start from 0, and the intensities in different conditions are only affected by the system factor. The spectrum was collected after the plasma was turned on for 1 minute.

RESULTS AND DISCUSSION

Emission spectra of intrinsic a-Si:H film deposition

Figure 1 shows the emission spectra of different intrinsic a-Si:H film deposition conditions. The SiH_4 is the only feeding gas and the flow rate is kept constant (50 sccm) in all conditions. There are several major spectral peaks, i.e., SiH^* (414 nm), H_α^* (656 nm), and H_β^* (486 nm). The intensities of SiH^*, H_α^*, H_β^*, and H_2 increase with the increase of power, e.g., 100 W vs. 200 W at SiH_4 50 sccm and 400 mT. However, when the power is low or the pressure is high, e.g. at 80 W and 150 mT or 100 W and 500 mT, only the SiH^* is detected. The SiH_4 plasma contains a large amount of atomic hydrogen that plays a dual role in the film formation process, i.e., to passivate the dangling bonds or to etch the weak Si-Si bond [9,10]. The low power or high pressure deposited film may contain many weak Si-Si and unsaturated dangling bonds, which affects the device performance. For example, the *p-i-n* solar cell prepared from the low power or high pressure *i*-layer has a lower conversion efficiency than those of cells with the *i*-layers deposited from other conditions. It was reported that the H_α^*/SiH^* intensity ratio was related to the crystalline phase formation in the PECVD μ-Si:H film [6]. When the H_2 concentration is over 97%, the crystalline phase formed [9]. Since the feed streams of Fig. 1 samples contain no H_2, the deposited films are amorphous as confirmed with the Raman spectroscope, i.e., only a broad feature near 480 cm^{-1} [9].

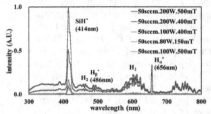

Figure 1. OES spectra for *i*- a-Si:H deposition under different conditions .

Figure 2 shows the deposition rate change with (a) the SiH^* intensity and (b) the H_β/H_α ratio. The dissociation of SiH_4 results in the created of the excited SiH^* [11]. The dissociation of H_2 can lead to the formation of H^* followed by H_α Balmer emission at 656 nm or H_β Balmer emission at 486 nm [12,13]. The H_β^*/H_α^* density ratio can be used to estimate the electron temperature T_e in the plasma phase [7]. The increase of the power promotes both the electron density and T_e [8]. The high electron density contributes to the generation of large amount of SiH^*. The high T_e means high kinetic energy of electrons, which favors the dissociation of molecules. In addition, influences of pressure on the SiH^* and H_β^*/H_α^* ratio are dependent on the magnitude of power. In the SiH_4 plasma, the dissociation and recombination mechanisms affect the film deposition phenomena [13]. With the increase of the pressure, the feed gas molecule's dissociation rate can increase, e.g., showing as the increase of the

SiH^* concentration, or decrease, e.g., showing as the increase of the recombination rate. For the low power case, e.g., 100 W, the latter is more pronounced than the former, which results in the decrease of the deposition rate with the increase of the pressure. For the high power case, e.g., 200 W, the former is more pronounced than the latter, which results in the opposite trend.

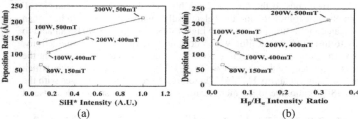

(a)   (b)

Figure 2. The dependence of deposition rate on (a) SiH^* intensity and (b) H_β^*/H_α^* ratio.

Emission spectra of n^+ a-Si:H film deposition

In the n^+ and p^+ a-Si:H deposition processes, the peaks of the dopant related free radicals could not be detected probably due to the low dopant gas concentration in the feed gas stream, e.g., 0.01%-0.13% for PH_3 and 0.15%-0.37% for B_2H_6. The SiH^*, H_α^*, H_β^*, and H_2 concentrations in the $SiH_4/H_2/PH_3$ and $SiH_4/H_2/B_2H_6$ plasmas are almost the same as those in the SiH_4/H_2 plasma. Figure 3(a) shows the power effect on the radicals intensities at $SiH_4/H_2/PH_3$ 60/1000/20 sccm, 800 mT, and 250°C. All intensities increase with the increase of power. For example, the SiH^* and H_α^* intensities at 500 W are 1.55 and 2 times those at 300W, separately. The increase of the H_α^* intensity is more obvious than the increase of the SiH^* intensity. H_2 can be contributed by the dissociated SiH_4 and the feed gas, but SiH^* can only be from of the dissociated SiH_4. Since the H_β^*/H_α^* ratio in Fig. 3(a) is independent of the power, the T_e is not influenced by the increase of power. Therefore, the increase of the power mainly contributes to the increase of the electron density. The high electron density can facilitate the dissociation of SiH_4, PH_3, and H_2, and thus accelerate the deposition rate, as shown in Figure 3(b).

Figure 4(a) shows the radicals concentrations vs. the n^+ film deposition rate. The samples were prepared under different process conditions, e.g., flow rates of SiH_4 (10-60 sccm), PH_3(6.88% in H_2) (1-20 sccm), and H_2 (1000 sccm), power (300-500 W), and pressure (750-850 mT). The deposition rate increases roughly with the increase of the SiH^* intensity or the decrease of the H_α^* intensity. There is a drastic increase of the deposition rate over a narrow window of the atomic hydrogen density probably due to the reduction of the etch effect on the film and the increase of the film precursor concentration, e.g. SiN_x. It is unclear currently whether the dopants affect the deposition rate. Since the PH_3 concentration in the feed stream is very low (0.01%-0.13%), the flow rate of doping gas should not contribute to such a large change of deposition rate [4]. Figure 4(b) shows the free radicals intensities vs. the n^+ film resistivity. The n^+ resistivity is determined by (1) the concentration of electrically active dopants, and (2) the morphology. The n^+ films produced in this experiment are all a-Si:H, i.e. the morphology of these films are similar. The resistivity increases with the increase of the H_α^* intensity but decreases with the increase of the SiH* intensity. Also, the resistivity increases drastically with the slight increase of the H$_\beta$* intensity but smoothly with the increase of the H$_\alpha$* intensity. It has been reported that when the PH$_3$/SiH$_4$ ratio in the feed stream is higher than 10^{-2}, the P incorporation saturates at 10^{-3} in the film [14]. The range of PH$_3$/SiH$_4$ ratio in Fig.4 is 0.014-0.023. Therefore, at the low SiH* and high H$_\alpha$* intensity conditions, which corresponds to the low SiH$_4$ and high H$_2$ dissociation efficiencies, P atoms are probably incorporated in the electrically inactive form, rather than substitutionally located in the film [4]. Also, the higher dissociation rate of H$_2$ can lead to higher possibility to form PH$_x$ in the film, which does not contribute to the conductivity. Although H atoms can passivate the dangling bond to increase the carrier lifetime, at the high concentration, it can lead to the coexistence of SiH, SiH$_2$ and SiH$_3$ in the film, which lowers the charge carrying capability [15, 16]. Therefore, the resistivity increases with the increase of the H$_\alpha$*/SiH* ratio due to the increase of the H$_\alpha$* intensity and the decrease of the SiH* intensity.

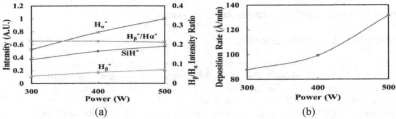

Figure 3. Power effects on (a) radicals intensities and (b) deposition rate. SiH$_4$/H$_2$/ PH$_3$ 60/1000/20 sccm, 800 mT.

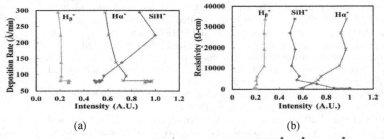

Figure 4. (a) deposition rate and (b) n^+ film resistivity vs. SiH*, H$_\alpha$*, and H$_\beta$* intensities.

Figure 5(a) shows that when the H$_\alpha$*/SiH* ratio is increased from 0.6 to 1.8, the resistivity increases almost 100 times. However, if the H$_\alpha$*/SiH* ratio is too low and the SiH* intensity is high, the dopant concentration in the film will be low. For example, for the i- layer deposited at SiH$_4$ 50 sccm, 100 W, and 500 mT, the H$_\alpha$*/SiH* ratio is about 0.015. The resistivity is too high for the 4-point probe measurement. It also shows that the film's resistivity is almost not affected by the H$_\beta$*/H$_\alpha$* ratio, i.e., the T$_e$. Figure 5(b) shows the film's resistivity and the H$_\alpha$/SiH* ratio vs. the SiH$_4$ flow rate at H$_2$/PH$_3$ 1000/20 sccm, 300 W, 800 mT, and 250°C. With the same PH3 flow rate, the resistivity decreases with the decrease of the H_α^*/SiH^* ratio because the SiH^* intensity increases faster than the H_α^* intensity does, leading to more electrically activated P dopants in the film, which verify the relationship between the H_α^*/SiH^* ratio and the resistivity in Fig. 5(a). Therefore, the low n^+ film resistivity can be obtained by adjusting the deposition parameters, such as power, pressure, and feed gas composition, toward the low H_α^*/SiH^* ratio. For example, within the deposition conditions in this study, the lowest resistivity of 426 Ω-cm was obtained under the condition of SiH_4/PH_3 (6.88% in H_2)/H_2 60 sccm/20 sccm/1000 sccm, 300 W, 800 mT.

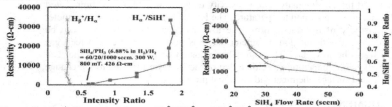

Figure 5. (a) n^+ film resistivity vs. H_α^*/SiH^* and H_β^*/H_α^* ratios, and (b) Effects of SiH_4 flow rate on H_α^*/SiH^* ratio and film resistivity. H_2/PH_3 1000/20 sccm, 300 W, 800 mT.

Emission spectra of p^+ film deposition

In the p^+ a-Si:H thin film deposition process, the relationships among deposition rate, resistivity, and free radicals intensities are similar to those in the n^+ a-Si:H thin film deposition. All intensities increase with the increase of power and the T_e does not change much with the increase of power. The deposition rate also increases with the increase of the SiH^* intensity as well as with the decrease of the H_α^* intensity. In addition, similar to the n^+ results, the p^+ film's resistivity increases with the increase of the H_α^*/SiH^* ratio due to the increase of H_α^* intensity and the decrease of the SiH^* intensity. Also, T_e does not affect the resistivity.

Dopant gas flow rate effect of n^+ and p^+ film deposition rates

Figure 6 shows the the dopant gas flow rate effect on deposition rates of (a) n^+ and (b) p^+ films. For the n^+ film deposition, the addition of a small amount of PH_3 in the feed stream increases the deposition rate. However, the deposition rate does not increase with the further increase of the PH_3 flow rate. The increase of the deposition rate may be due to the change of the electrical and chemical properties of the film similar to that occured in the growth of the P implanted epitaxy layer [17]. For the p^+ film deposition, the deposition rate increases with the increase of the B_2H_6 flow rate, which may be due to the reduction of

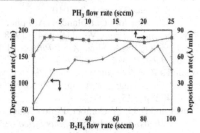

Figure 6. Deposition rate vs. PH_3 (6.88% in H_2) flow rate in n^+ deposition rate and B_2H_6 (2% in H_2) flow rate in p^+ deposition rate.

the H_2 desorption barrier, enhancement of the Si adsorption rate, or the increase of the SiH_x sticking coefficient [18,19].

SUMMARY

The PECVD i-, n^+, and p^+ a-Si:H thin film deposition processes have been studied using the OES to monitor changes of the SiH^*, H_α^*, and H_β^* intensities. For all films, the deposition rate increases with the increase of the SiH^* intensity but decreased with the increase of the H_α^* intensity. The resistivity of the doped film is a complicated function of the electron energy and temperature. The H_α^*/SiH^* ratio is a critical parameter affecting the resistivity of the doped film. The low resistivity of doping film can be obtained by adjusting the deposition parameter, such as power, pressure, and feed gas composition, toward the low H_α^*/SiH^* ratio. The inclusion of PH_3 or B_2H_6 in the feed stream gives a higher film deposition rate than that of the intrinsic film due to the change of the film's bulk and surface chemical and electrical properties.

ACKNOWLEDGMENTS

Authors acknowledge the support of this research by NSF CMMI 0968862 project.

REFERENCE

1. J. N. Bullock and C. H. Wu, *J. Appl. Phys.*, Vol. 69, p. 1041, 1991
2. D. E. Carlson and C. R. Wronsky, *Appl. Phys. Lett.*, Vol. 29, p. 602, 1976
3. Y. Kuo, *MRS Proc.*, Vol. 282, p. 197, 1985
4. Y. Kuo, *Appl. Phys. Lett.*, Vol. 71, p. 2821, 1997
5. Tochikubo F, Suzuki A, Kakuta S, Terazono Y and Makabe T, *Appl. Phys.*, Vol.68, p. 5532, 1990
6. H. Jia, J.K. Saha, N. Ohse, and H. Shirai, *J. Non-Cryst. Solids*, Vol. 352, p. 896, 2006.
7. K. Saito and M. Kondo, *Phys. Status Solidi A*, Vol. 207, No. 3, p. 535, 2010.
8. S. K. Ram, L. Kroely, S. Kasouit, P. Bulkin, and P. R. Cabarrocas, *Phys. Status Solidi C*, Vol. 7, No. 3-4, p. 553, 2010
9. C. C. Tsai, R. Thompson, C. Doland, F. A. Ponce, G. B. Anderson, and B. Wacker, *Mat. Res. Soc. Symp. Proc.*, Vol. 118, p. 49, 1988
10. U. Kroll, J. Meier, and A. Shah, *Appl. Phys.*, Vol. 80, p. 4971, 1996.
11. A. Matsuda, *J.J.A.P.*, Vol. 43, p. 7909, 2004
12. J. Perrin and J.P.M. Schmitt, *Chemical Physics*, Vol. 67, p.167, 1982
13. H. Yang, C. Wu, J. Huang, R. Ding, Y. Zhao, X. Geng, and S. Xiong, *Thin Solid Films*, Vol. 472, p. 125, 2005
14. W. E. Spear and P. G. Le Comber, Philos. Mag., Vol. 33, p. 935, 1976.
15. R. A. Street, *Hydrogenated Amorphous Silicon*, Cambridge University Press, 1992.
16. M. H. Brodsky, M. Cardona, and J. J. Cuomo, *Phys. Rev. B*, Vol. 16, p. 3556, 1997.
17. H. Nominanda and Y. Kuo, *Electrochem. Soc. Proc.*, Vol. 17, p. 1-9, 2002.
18. P. J. Hay, R. C. Boehm, J. D. Kress, R. L. Martin, Surf. Sci., Vol. 436, p. 175, 1999
19. H. Nominanda and Y. Kuo, *Electrochem. Soc. Proc.*, Vol. 17, p. 1-9, 2002.

Mater. Res. Soc. Symp. Proc. Vol. 1536 © 2013 Materials Research Society
DOI: 10.1557/opl.2013.753

On the Origin of the Urbach Rule and the Urbach Focus

J. A. Guerra[1,2], L. Montañez[1], F. De Zela[1], A. Winnacker[2] and R. Weingärtner[1,2]

[1]Pontifical Catholic University of Peru, Sciences Department, Physics Section, Av. Universitaria 1801, Lima 32, Peru.
[2]University of Erlangen-Nurnberg, Institute of Material Science 6, Martensstr. 7, 91058 Erlangen, Germany

ABSTRACT

A simple derivation of sub-bandgap exponential tails and fundamental absorption equations ruling the optical absorption of amorphous semiconductors are presented following the frozen phonon model. We use the Kubo-Greenwood formula to describe the average transition rate for the optical absorption process. Asymptotic analysis leads to the commonly observed exponential tail as well as the Tauc expression for the fundamental absorption. We test our theoretical results with experimental absorption coefficients of amorphous Si:H, SiC:H, AlN and SiN. The validity of the Urbach focus concept is evaluated.

INTRODUCTION

Crystalline wide bandgap semiconductors have been of increasing interest in the past decade due to their advantageous properties for applications in the visible and ultraviolet regions. Their amorphous counterparts share such features [1-5] and offer additional properties such as low production costs and higher doping incorporation, which makes them attractive for device applications. Thus, in order to design devices with these amorphous materials a good understanding of their optical properties is necessary. One important aspect is the wavelength dependent absorption coefficient in the fundamental and adjacent region: The Tauc region describes the extended states related transitions. The Urbach region accounts for transitions related to localized states, observed as the so called exponential tail in the absorption coefficient. Tauc derived the behavior of the fundamental absorption following a well-known theoretical approach used for crystals. Here indirect transitions are allowed for simulating energy conservation by disorder, i.e. by the "frozen phonons". However, this method fails in the Urbach region. The Toyazawa's group [6,7] introduced the static disorder effect assuming a Gaussian distribution for thereby random site energies into the electronic Hamiltonian, achieving the exponential behavior of the band edges. Other contributions, such as the work of Dunstant [8], Soukoulis, Sajeev, Cohen and Economou [9, 10] and later on the contribution of the group of O'Leary on understanding the electronic properties of amorphous materials [11-14], introduced the effect of thermal fluctuations in the band edge (or the band edge fluctuations due to static disorder) and calculated the average absorption coefficient for both the Tauc and the Urbach regimes.

Here we propose a simple, analytical treatment which accounts for the Tauc behavior and the exponential tail that has been generally observed in the absorption coefficient of amorphous semiconductors. Our treatment is based on the formalism employed for the transition rate calculations in the one electron approximation at finite temperature (for crystals) giving a new meaning to the Urbach focus.

THEORETICAL APPROACH

It is well established that the origin of the band tails in the absorption coefficient of amorphous semiconductors lays in the topological disorder and thermal vibrations [6-16]. The latter is normally small and insignificant at low temperatures. However, the fact that both have the same functional behavior gives us the possibility to model both effects by thermal vibrations alone. The physical reason lays in the assumption that a disordered material may be considered as an ordered one at a high (fictive) temperature frozen in time (frozen phonon model). In order to calculate the absorption coefficient in such systems we need to calculate the average electronic transition rate R_{cv} between the conduction and valence bands due to the optical absorption process of an ensemble of electrons in a fictive heat bath. This can be achieved by extending the one electron approximation considering the electronic occupation degree and the stimulated emission process (see equation 1).

$$\langle R_{cv} \rangle = \frac{2\pi}{\hbar} \left(\frac{\tilde{E}e}{2\omega m_e} \right)^2 \sum_{k_c, k_v} |M_{cv}|^2 [f(E_v)]\{\delta(E_c - E_v - \hbar\omega)$$
$$-\delta(E_c - E_v + \hbar\omega)\}[1 - f(E_c)]\delta_{k_c, k_v} \qquad (1)$$

Here, $\tilde{E}$ is the electric field of the incident. M_{cv} is the electronic transition matrix element between the conduction and valence bands. $f(E)$ is the Fermi distribution. m_e is the free electron mass, and e is the electron charge. Equation 1 can be reduced to the Kubo-Greenwood formula, which is commonly used to describe the temperature dependence of the electrical conductivity of semiconductors [17,18]. Here it describes the disorder dependence of the optical absorption process, since we may interpret the temperature T in the Fermi distribution function as a measure of disorder.

$$\langle R_{cv} \rangle = \frac{2\pi}{\hbar} \left(\frac{\tilde{E}e}{2\omega m_e} \right)^2 \sum_{k_c, k_v} |M_{cv}|^2 \{f(E_v) - f(E_c)\}\delta(E_c - E_v - \hbar\omega)\delta_{k_c, k_v} \qquad (2)$$

We follow Tauc's calculation for amorphous semiconductors [19], i.e., relaxing the conservation of k and writing $\langle R_{cv} \rangle$ in its integral form, by using the valence $D_v(E_v)$ and conduction $D_c(E_c)$ electronic density of states in the free electron approximation. The

140

absorption coefficient is obtained as shown in equation 3, where the energy reference is taken so that $E_c(0) = E_g$ and $E_v(0) = 0$.

The absorption coefficient is given by

$$\alpha = \frac{\alpha_0}{\hbar\omega} \int_{E_g}^{\hbar\omega} \int_{E_0-\hbar\omega}^{0} \sqrt{E_c - E_g} \sqrt{-E_v} \frac{df(E_c - E_v - \hbar\omega)}{dE_v} dE_v dE_c \qquad (3)$$

with

$$\alpha_0 = \frac{2}{cn\varepsilon_0} \left(\frac{e}{m_0}\right)^2 \frac{(m_e^* m_h^*)^{3/2}}{\pi^2 \hbar^3} |M_{cv}|^2.$$

Here, m_e^* is the electron effective mass, m_h^* is the hole effective mass, n is the refractive index, ε_0 is the vacuum electrical permittivity and c is the vacuum speed of light.

Equation 3 can be written more compactly as equation 4 using the Di-Logarithm function of x $\text{Li}_2(x)$:

$$\alpha(\hbar\omega) = -\frac{\pi}{4} \frac{\alpha_0}{\beta^2 \hbar\omega} \text{Li}_2\left(-e^{\beta(\hbar\omega - E_0)}\right) \qquad (4)$$

E_0 is defined as $E_0 = E_g - \mu$, and μ plays the role of a pseudo-chemical potential that depends on the matrix doping. $\beta = 1/(k_B T)$ is the definition of the Urbach slope. Notice that μ enters only as a shift of the bandgap and therefore it can be interpreted as the bandgap shift due to the doping effect or other mechanisms that may vary the band edge.

Asymptotic analysis of equation 4 leads to the Urbach and Tauc expressions, respectively. Mathematically, we obtain the same coefficient α_0 in both regimes. Physically, we allow use different coefficients, since the nature of the transitions is different on account of different matrix elements $|M_{cv}|$ (see equation 5). In this work we focus only on the Urbach expression. The exponential tail described does not predict a constant Urbach focus, on the contrary it predicts a whole region (see figure 1).

$$\alpha(\hbar\omega) \approx \frac{\pi}{8} \frac{\alpha_0}{\hbar\omega} \begin{cases} \frac{2}{\beta^2} e^{\beta(\hbar\omega - E_0)} & , \quad \hbar\omega < E_0 \\ (\hbar\omega - E_0)^2 + \frac{\pi^2}{3\beta^2} & , \quad \hbar\omega > E_0 \end{cases} \qquad (5)$$

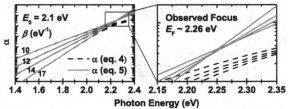

Figure 1. Absorption coefficient via the equation 4 and 5 for $\hbar\omega \ll E_0$, using the shown parameters. The Urbach focus region is enlarged to appreciate it better.

The average focus is located at $E_F = E_0 + 2(\log(\beta)/\beta)$. The wideness of the focus region depends on how large the spreading of β is. In our model the constant E_0 is the bandgap in the absence of the disorder plus a shift due to the doping or other effects that my change the band edges.

RESULTS AND DISCUSSION

To test our model and to compare it with the Urbach rule a straightforward procedure has been followed. F. Orapunt and S. O'Leary [20] showed that the Urbach focus value of a-Si:H found by Cody [15] is in agreement with independent fits of the former set of absorption coefficients data using the Urbach rule (equation 6) and hence without enforcing the existence of the focus through a global fit [21]. The fits are performed in order to find the parameters C_F and β for each absorption coefficient independently. Subsequently, the existence of the focus can be tested if the linear relation $\log(C_F) = \log(\alpha_F) - \beta E_F$ is observed. This analysis allows testing the existence of either the average focus E_F or the bandgap related parameter E_0 when extended to our model (equation 7).

$$\alpha(\hbar\omega) = C_F e^{\beta(\hbar\omega)} \leftrightarrow C_F = \alpha_F e^{-\beta E_F} \tag{6}$$

$$\alpha(\hbar\omega) = \frac{\pi}{4} \frac{C_0}{\hbar\omega\beta^2} e^{\beta(\hbar\omega)} \leftrightarrow C_0 = \alpha_0 e^{-\beta E_0} \tag{7}$$

Both types of fits are shown in figure 2 using the Urbach rule (equation 6) for the absorption coefficient of a-Si:H taken from [15]. The Global fit is performed in such a way that the parameters α_F and E_F are shared by the set of absorption coefficients.

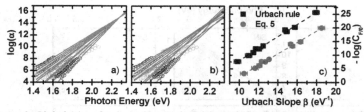

Figure 2. *Global (a) and independent (b) fits of the Urbach region of the absorption coefficient of a-Si:H using the Urbach rule. $log(C_{F/0})$ vs β using the parameters $C_{F/0}$ and β obtained from independent fits using the equations 6 and 7 (c). The analysis in both cases predicts the respective constants: $E_0 = 2.08 \pm 0.06$ eV and $E_F = 2.21 \pm 0.06$ eV in agreement with the corresponding global fits.*

The linear fits depicted in figure 2.c show that both fitting procedures are consistent with the notion of E_F and E_0 in the case of a-Si:H. Nevertheless, while the Urbach rule predicts E_F empirically, the latter predicts the bandgap in the absence of disorder plus a shift due to the doping effect E_0 and the Urbach focus E_F as being actually a small region instead of a constant.

The existence of the Urbach focus, or at least a focus region, has been reported in a series of materials, e.g. a-Ge [22], LnSe [23], a-SiC [21,24], and a-AlN [21]. Here, we test both models for the case of the wide bandgap amorphous semiconductors AlN, SiN and SiC:H. These materials were grown by reactive RF magnetron sputtering from high purity targets and were annealed at different temperatures for 15 min in an inert atmosphere at low pressure in order to vary their Urbach energy. More details on the deposition and thermal treatment of these samples can be found in [21,25]. The absorption coefficient of these materials exhibits an Urbach focus region (see figure 3). It was measured through optical transmittance spectroscopy.

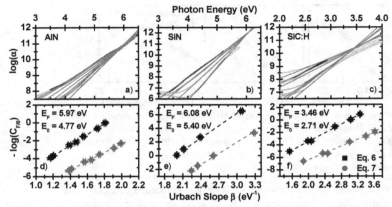

Figure 3. *Global fit using the equation 7 showing the Urbach focus region in each material (a), (b) and (c). The respective analysis in order to validate the existence of the constants E_F and E_0 (d), (e) and (f).*

In summary, a general equation that describes the whole range of interest of the absorption coefficient (Urbach and Tauc regions) observed in amorphous semiconductors has been presented. It predicts the value of the bandgap in the absence of disorder. The proposed model covers the concept of the Urbach focus and is in agreement with it as a region of slightly deferring Urbach foci.

ACKNOWLEDGMENTS

This research was funded by the General Research Direction (DGI) of the Pontifical Catholic University of Peru (PUCP). The authors have been supported by the PUCP under the PhD scholarship program *Huiracocha* (J. A. Guerra) and by the German Academic Exchange Service (DAAD) under the contract number D11-15283 (R. Weingärtner).

REFERENCES

1. F. Zhu , J. Hu, I. Matulionis,T. Deutsch, N. Gaillard, E. Miller, and A. Madan, *Solar Energy Book*, chapter 15, (Vukovar: Intech, 2010).
2. R. Gerhardt, *Properties and Applications of Silicon Carbide*, (Vienna: Intech, 2011).
3. H. H. Richardson, P. G. Van Patten, D. R. Richardson, and M. E. Kordesch, *Appl. Phys. Lett.* **80**, 2207 (2002).
4. D. Adachi, R. Kitaike, J. Ota, T. Toyama, and H. Okamoto, *J. Mater. Sci. Mater Electron.* **18**, S71 (2007).

5. A. Wakahara, K. Takemoto, F. Oikawa, H. Okada, T. Ohshima, and H. Itoh, *Phys. Stat. Sol. A* **205**, 56 (2008).
6. H. Sumi, and Y. Toyazawa, *J. Phys. Soc. Jpn.* **31**, 342 (1971).
7. S. Abe, and Y. Toyazawa, *J. Phys. Soc. Jpn.* **50**, 2185 (1981).
8. D. J. Dunstan, *J. Phys. C: Solid State Phys.* **16**, L567-L571 (1983).
9. C. M. Soukoulis, M. H. Cohen, and E. N. Economou, *Phys. Rev. Lett.* **53**, 616 (1984).
10. J. Sajeev, C. M. Soukoulis, M. H. Cohen, and E. N. Economou, *Phys. Rev Lett.* **57**, 1777 (1986).
11. S. K. O'Leary, S. Zukotynski, and J. M. Perz, *Phys. Rev. B* **51**, 4143 (1995).
12. S. K. O'Leary and P. K. Lim, *Solid State Communications* **104**, 17 (1997).
13. S. K. O'Leary, *Appl. Phys. Lett.* **72**, 1332 (1998).
14. T. H. Nguyen, and S. K. O'Leary, *J. Appl. Phys.* **88**, 3479 (2000).
15. G. D. Cody, T. Tiedje, B. Abeles, B. Brooks, and Y. Goldstein, *Phys. Rev. Lett.* **47**, 1480 (1981).
16. M. Letz, A. Gottwald, M. Richter, V. Liberman, and L. Parthier, *Phys. Rev. B* **81**, 155109 (2010).
17. N. F. Mott, and E. A. Davis, *Electronic processes in non-crystalline materials* Second edition (Oxford University Press, 1979).
18. L. L. Moseley, and T. Lukes, *Am. J. Phys.* **46**, 676 (1978).
19. J. Tauc, *Mat. Res. Bull.* **3**, 37 (1968).
20. F. Orapunt, and S. K. O'Leary, *Appl. Phys. Lett.* **72**, 1332 (2004).
21. J. A. Guerra, L .Montañez, O. Erlenbach, G. Galvez, F. De Zela, A. Winnacker, and R. Weingärtner, *J. Phys. Conf. Ser.* **274**, 012113 (2011).
22. G. D. Cody, *Mater. Res. Soc. Symp. Proc.* **862**, A1.3.1 (2005).
23. B. Abay, H. S. Güder, H. Efeoglu, and Y. K. Yogurtçu, *J. Phys. D: Appl. Phys.* **32**, 2941 (1999).
24. F. Zhang, H. Xue, Z. Song, Y. Guo, and G. Chen, *Phys. Rev. B* **46**, 4590 (1992).
25. R Weingärtner, J. A. Guerra, O. Erlenbach, G. Gálvez, F. De Zela, and A. Winnacker, *Mat. Sci. and Eng. B* **174**, 114 (2010).

Mater. Res. Soc. Symp. Proc. Vol. 1536 © 2013 Materials Research Society
DOI: 10.1557/opl.2013.909

Ultrafast Optical Measurements of Acoustic Phonon Attenuation in Amorphous and Nanocrystalline Silicon

Joseph A. Andrade,[1] Bryan E. Rachmilowitz,[1] Brian C. Daly,[1] Theodore B. Norris,[2] B. Yan,[3] J. Yang,[3] and S. Guha[3]

[1] Physics and Astronomy Department, Vassar College, 124 Raymond Ave., Poughkeepsie, NY 12604, USA

[2] Department of Electrical Engineering and Computer Science, University of Michigan, Ann Arbor, MI 48109, USA

[3] Uni-Solar Ovonic LLC, Troy, MI 48084, USA

ABSTRACT

We have measured the attenuation of longitudinal acoustic waves in a series of amorphous and nanocrystalline silicon films using picosecond ultrasonics. We determined the attenuation of amorphous Si to be lower than what is predicted by theories based on anharmonic interactions of the ultrasound wave with localized phonons or extended resonant modes. We determined the attenuation of nanocrystalline Si to be nearly one order of magnitude higher than amorphous Si.

INTRODUCTION

Amorphous and nanocrystalline hydrogenated silicon thin films (a-Si:H and nc-Si:H) are prevalent in many devices, including solar cells. In an effort to extract a more complete picture of the thermal and mechanical properties of such important materials, we have measured the attenuation of very high frequency ultrasound (50 – 100 GHz) in these Si layers using an ultrafast optical technique known as picosecond ultrasonics. The attenuation of high frequency sound waves in amorphous and nanocrystalline materials is not yet well understood. This is not surprising considering the uncertainty that remains in the study of the thermal conductivity of these materials, a topic that is directly related. In a recent paper [1] we reported the first measurements of ultrasound attenuation in a-Si:H in the ~100 GHz regime and we extend that work to include measurements of nc-Si:H here.

Several experimental [2,3] and theoretical [4,5] studies have been made of the thermal conductivity of a-Si (not hydrogenated) and a-Si:H including recent reports of anomalously high thermal conduction in hot wire chemical vapor deposited (HWCVD) samples [6,7]. In all cases, analysis of the thermal conductivity measurements typically boils down to an attempt to determine the overall magnitude and frequency dependence of the lifetime of long wavelength (~ 1 THz) acoustic phonons or localized phonon-like excitations. It is envisioned that the accurate measurements of attenuation in the sub-THz regime that are possible using the technique of picosecond ultrasonics will assist in the overall picture of thermal transport in amorphous materials. In addition to these fundamental issues, we also envision that picosecond ultrasonics could become a valuable diagnostic tool for determining interface quality and pore/void density in manufactured amorphous and polycrystalline silicon layers.

EXPERIMENTAL DETAILS

The a-Si:H and nc-Si:H thin films were grown using a modified VHF glow discharge method at a high deposition rate on polished steel substrates. The deposition conditions were similar to that used in the deposition of high-efficiency solar cells [8]. A proper hydrogen dilution with an H_2 and SiH_4 mixture was used to improve material quality. A series of samples with different thicknesses was made under the same conditions (temperature, gas flow rates, pressures, VHF power). Only the time was varied to achieve desired film thicknesses. For the nc-Si:H samples, dilution profiling was used to control crystallinity along the growth direction [9]. These nc-Si:H layers used in this study were deposited under the condition of best nc-Si:H solar cell performance. Raman and cross-sectional TEM measurements showed roughly 50% crystalline volume fraction. XRD showed a (220) preferential orientation with a grain size of around 30 nm. Thin Al layers for use as transducers in the optical experiment were deposited on the top surface of the Si layers using a thermal evaporation method.

To measure the attenuation we used the ultrafast optical pump-probe technique known as picosecond ultrasonics [10,11], and specifically followed the methods described by Morath and Maris [12]. A diode laser–pumped Ti:sapphire oscillator (Coherent Mira) that operates at a repetition rate of 76 MHz and emits pulses that are ‿100 fs in duration was used to perform the experiment. Pump and probe beams with average power ~10 mW were focused down to the same ~15-μm diameter spot on the Al-coated silicon film samples. The absorption of pump pulses caused rapid thermal expansion of the Al transducer layer, which launched longitudinal acoustic strain pulses into the Si film. The acoustic pulses are roughly single cycle with the compressive strain leading the rarefacting strain, and they contain a broad range of frequencies of coherent long-wavelength acoustic phonons. The frequencies generated in an arbitrarily thick Al layer would be determined by the optical absorption depth of the 800 nm light, which can vary up to several 10's of nm for sputtered Al films. For thinner films, in which the light penetrates the full thickness of the film, the thickness itself plays a role in determining the bandwidth of frequencies generated, with thinner films yielding higher frequencies. We used Al transducer thicknesses of 30 and 15 nm in order to obtain variation of the central acoustic frequency. The 30 nm films provided measureable frequencies in the range of 30 – 90 GHz while the 15 nm films provided measureable frequencies ranging from 40 – 120 GHz. The strain pulses were detected by time-delayed optical probe pulses that are sensitive to changes in the reflectivity ΔR of the Al film.

Figure 1 shows ΔR versus delay time for three a-Si:H samples and two nc-Si:H samples with the varying thicknesses as labeled. For this graph, the initial electronic response of the Al film near 0 ps has been removed for all five sets of data, as has the thermal decay of the reflectivity change. The signals that remain for each data set represent the detection of two acoustic pulses: one that has traveled a complete round trip through the Si film and back to the Al transducer after reflecting from the steel substrate and one that has made two such round trips. The time delay between the two signals can be used to accurately measure the thickness of the Si film if a value of the sound velocity is assumed. For visual clarity, the five data sets have all been normalized to have the same amplitude of the first signal $\Delta R_1(t)$ so one can get a sense of the reduction in amplitude of the second signal $\Delta R_2(t)$ in each case. Two effects are immediately observable from this graph. First, the overall attenuation of the signal is significantly affected by increasing the thickness of the samples. Second, the attenuation is much larger in the nc-Si:H, as despite using significantly thinner samples in those cases, we observe a comparable reduction in the size of $\Delta R_2(t)$.

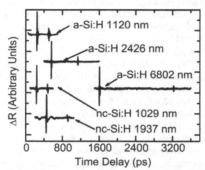

Figure 1. Change in reflectivity ΔR versus delay time for 3 a-Si:H and 2 nc-Si:H samples with 15 nm Al transducer thicknesses. The data are offset for clarity and the thermal background has been subtracted from each data set. (Reproduced from Ref. 1)

To obtain quantitative results for the attenuation α as a function of the frequency ω, we use the expression:

$$\alpha = \frac{1}{d}\ln\left(\frac{r\Delta R_1(\omega)}{\Delta R_2(\omega)}\right) \tag{1}$$

where d is the round trip distance through the film and r is an experimentally determined reflection coefficient. r accounts for all losses at the Si-steel interface, as some part of the acoustic wave will propagate into the steel rather than return to the free surface to be detected by the optical probe pulses. In addition, there may be losses due to imperfections at the Si-steel interface or the Al-Si interface. We determine r by measuring the $\Delta R_1(\omega) / \Delta R_2(\omega)$ for our thinnest samples, where the interface loss dominates over the loss that is intrinsic to the Si layer. We find r to be 0.36 for both the a-Si:H and nc-Si:H samples. As a comparison, we can also calculate a value of r from the acoustic mismatch between the Si film and the steel using the expression $r = (\rho_{Steel}v_{Steel} - \rho_{Si}v_{Si}) / (\rho_{Steel}v_{Steel} + \rho_{Si}v_{Si})$. Using literature values of $\rho_{Steel} = 7.9$ g/cm^3, $v_{Steel} = 5790$ m/s [13], $\rho_{Si} = 2.3$ g/cm^3 [14], and $v_{Si} = 8400$ m/s [15] we find $r = 0.41$, just slightly larger than our experimentally determined value. For a-Si:H samples thicker than 4000 nm and for nc-Si:H samples thicker than 1000 nm however, we find that there is a significant increase in the attenuation such that the component of attenuation that is intrinsic to the silicon itself can clearly be measured.

DISCUSSION

Using Eq. 1 and values of $\Delta R_1(\omega) / \Delta R_2(\omega)$ from 4072 nm and 6800 nm thick a-Si:H films we find $\alpha = 340 \pm 120$ cm^{-1} at 50 GHz and 780 ± 160 cm^{-1} at 100 GHz. The uncertainties were determined from repeated measurements made using the 15 nm and 30 nm transducer samples, and include the uncertainty in the experimental value of r. The values are about an order of magnitude smaller than those reported for other amorphous materials [12] including a-SiO$_2$ which was most recently reported by Devos et. al to have an attenuation of 18000 cm^{-1} at 219 GHz [16]. That the amorphous phase of Si should have a lower GHz attenuation than many other amorphous materials is not entirely surprising, as the 2002 review of low temperature thermal conductivity and internal friction by Pohl, Liu, and Thompson certainly indicates that the fourfold coordinated materials tend to demonstrate weaker anharmonicity [17]. Our measured values are higher than recently reported values for crystalline Si (74 cm^{-1} at 20 GHz, 87 cm^{-1} at 50 GHz, and 119 cm^{-1} at 100 GHz) [18,19], but are within an order of magnitude. While our measured attenuation indicates a linear dependence on frequency, the uncertainties are such that a quadratic dependence is by no means ruled out by our measurement.

In Fig. 2 we plot our measured values for the attenuation along with predictions for a-Si from the work of Fabian and Allen [4]. In that work, the authors appealed to a model of attenuation based on the work of Akhiezer [20], in which the sound wave causes the local thermal vibrations (phonons or phonon-like "vibrons") to be displaced from equilibrium. The sound wave loses energy as these vibrational states return to their original occupation numbers. They used a 3-D 1000 atom simulation employing realistic atomic coordinates and interatomic potentials in order to calculate the mode Gruneisen parameters to be used in a relaxation time approximation calculation [21]. Their calculations for the attenuation of ultrasound in a-Si are represented in Fig. 2 and are in reasonable agreement with our results. The complete model of Fabian and Allen includes the effect of the internal strain caused by the randomized atomic positions in the amorphous model. The presence of the internal strain in turn causes some low

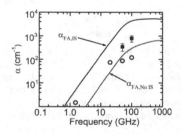

Figure 2. Measured attenuation versus frequency for a-Si:H (■). Literature values for attenuation in c-Si (○) as reported in Refs. 18,19. $\alpha_{FA,IS}$ and $\alpha_{FA,No\ IS}$ are from Ref. 4 and represent the attenuation calculated with and without internal strain. (Reproduced from Ref. 1)

frequency resonant modes in softer, undercoordinated regions of the amorphous Si network to have very large negative Gruneisen parameters (as low as -30) [22], and these large values are responsible for much of the attenuation. Their model overestimates our measured attenuation by about a factor of 6 at both 50 and 100 GHz. Also plotted in Fig. 2 is the calculation of Ref. 4 when the internal strain effect is eliminated from the simulation. Our measured values fall directly between the values calculated with and without internal strain. It is interesting to note that better agreement appears to occur between the calculation without internal strain and the published values for the attenuation of crystalline Si [18,19,23], despite the random network structure of the numerical model.

In attempting to compare the theoretical predictions of Ref. 4 with our data, a few limitations must be noted. First, the calculation ignores the contribution of three-phonon scattering processes that directly involve the ultrasonic wave (a very low frequency phonon mode) under study, and therefore 100 GHz is expected to be the upper limit of that model's predictive power. Second, the model calculation was performed for a-Si, not a-Si:H, and the hydrogen content for our samples is estimated at 15%. While thermal conductivity results have been shown to have no significant dependence on H content from 1% up to 20% [2], it is possible that clusters of H atoms may have an impact on the coordination of the amorphous Si network, contributing to changes in the Gruneisen parameters discussed above. Future measurements of attenuation of a-Si:H with varying H content should be able to shed light on this issue. Lastly, we point out that while theoretical models of an amorphous material can provide a truly random network, a recent study of electron nanodiffraction patterns from a sputtered a-Si sample indicated that much of that sample was in fact composed of Si crystallites of $1 - 2$ nm [24].

Figure 3 shows the measured values for the attenuation in the nc-Si:H samples along with our data from the a-Si:H samples. We find that α in nc-Si:H is one order of magnitude larger than a-Si:H in this frequency range: $\alpha = 2990 \pm 630$ cm^{-1} at 40 GHz, 3770 ± 1320 cm^{-1} at 60 GHz, $\alpha = 5970 \pm 1400$ cm^{-1} at 80 GHz, and 7540 ± 660 cm^{-1} at 100 GHz. While larger scattering is not surprising given the comparable size of the nanocrystallites and the ultrasound wavelength, it is interesting to note that the material with the higher thermal conductivity exhibits the larger ultrasonic attenuation here even though we are only one order of magnitude lower in frequency than the long wavelength acoustic phonons that are largely responsible for heat transport. While the literature does not yet extend into the GHz regime, the attenuation of ultrasound in the MHz range in polycrystalline media and its dependence on grain size and frequency have been thoroughly reviewed [25]. In Figure 3 the two plotted lines are the result of calculations that use the average crystallite size, the ultrasound frequency, the material density, and the elastic moduli; the first assumes that we are in the Rayleigh regime ($\lambda \gg D$) and the second assumes the intermediate or stochastic regime ($\lambda \sim D$). Given that the wavelengths of the ultrasound in Si from $50 - 100$ GHz ranges from 170 nm - 85 nm, and the crystallite size D is 30 nm, the measurement resides somewhere between the two regimes. While the order of magnitude appears to agree more strongly with the Rayleigh scattering model, the frequency dependence appears to be considerably weaker than the expected fourth power, and it instead more readily agrees with the second power that is characteristic of the intermediate regime.

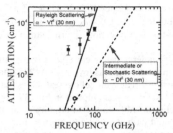

Figure 3. Measured attenuation versus frequency for nc-Si:H (■) and a-Si:H(○)

CONCLUSIONS

We have used picosecond ultrasonics to study the attenuation of high frequency ultrasound in a-Si:H and nc-Si:H. We find that the attenuation in a-Si:H is much lower than other amorphous materials and that the attenuation in nc-Si:H is about one order of magnitude higher than in a-Si:H.

ACKNOWLEDGEMENTS

The authors acknowledge the support of NSF Award DMR-1206681 entitled "RUI:Ultrafast Acoustics and Heat Transport in Nanostructures, Thin Films, and Crystals."

REFERENCES

1. D.B. Hondongwa, B.C. Daly, T.B. Norris, B. Yan, J. Yang, and S. Guha, Phys. Rev. B **83**, 121303R (2011).
2. D.G. Cahill, M. Katiyar, and J.R. Abelson, Phys. Rev. B **50**, 6077 (1994).
3. B.L. Zink, R. Pietri, and F. Hellman, Phys. Rev. Lett. **96**, 055902 (2006).
4. J. Fabian and P.B. Allen, Phys. Rev. Lett. **82**, 1478 (1999).
5. P.B. Allen and J.L. Feldman, Phys. Rev. Lett. **62**, 645 (1989).
6. X. Liu, J.L. Feldman, D.G. Cahill, R.S. Crandall, N. Bernstein, D.M. Photiadis, M.J. Mehl, D.A. Papaconstantopolous, Phys. Rev. Lett. **102**, 035901 (2009).
7. H.S. Yang, D.G. Cahill, X. Liu, J.L. Feldman, R.S. Crandall, B.A. Sperling, and J.R. Abelson, Phys. Rev. B. **81**, 104203 (2010).
8. G. Yue, B. Yan, J. Yang, and S. Guha, Mater. Res. Soc. Symp. Proc. **989**, 359 (2007).
9. B. Yan, G. Yue, J. Yang, S. Guha, D. L. Williamson, D. Han, and C.-S. Jiang, Appl. Phys. Lett. **85**, 1955 (2004).
10. C. Thomsen, H.T. Grahn, H.J. Maris, and J. Tauc, Phys. Rev. B **34**, 4132 (1986).
11. G.A. Antonelli, B. Perrin, B.C. Daly, and D.G. Cahill, MRS Bull. **31**, 607 (2006).
12. C.J. Morath and H.J. Maris, Phys. Rev. B. **54**, 203 (1996).
13. Handbook of Chemistry and Physics, 78th Ed. CRC Press (1997).
14. D.L. Williamson, Sol. Energ. Mat. Sol. C. **78**, 41 (2003).

15. W. Senn, G. Winterling, and M. Grimsdtch, in Physics of Semiconductors 1978, Ed. B.L.H. Wilson, Institute of Physics London 1979 p.709.
16. A. Devos, M. Foret, S. Ayrinhac, P. Emery, and B. Rufflé, Phys. Rev. B **77**, 100201(R) (2008).
17. R.O. Pohl, X. Liu, and E. Thompson, Rev. Mod. Phys. **74**, 991 (2002).
18. B.C. Daly, K. Kang, Y. Wang, D.G. Cahill, Phys. Rev. B **80**, 174112 (2009).
19. J.-Y. Duquesne and B. Perrin, Phys. Rev. B **68**, 134205 (2003).
20. A. Akhieser, J. Phys. (USSR) **1**, 277 (1939).
21. H. J. Maris, in *Physical Acoustics*, edited by W. P. Mason and R.N. Thurston (Academic, New York, 1971), Vol. 8.
22. J. Fabian and P.B. Allen, Phys. Rev. Lett. **79**, 1885 (1997).
23. A. A. Bulgakov, V. V. Tarakanov, and A. N. Chernets, Sov. Phys. Solid State **15**, 1280 (1973).
24. J.M. Gibson, M.M.J. Treacy, T. Sun, and N.J. Zaluzec, Phys. Rev. Lett. **105**, 125504 (2010).
25. E. P. Papadakis, J. Acoust. Soc. Am. **37**, 703 (1965).

Mater. Res. Soc. Symp. Proc. Vol. 1536 © 2013 Materials Research Society
DOI: 10.1557/opl.2013.910

Stress Analysis of Free-Standing Silicon Oxide Films Using Optical Interference

Imen Rezadad*, Javaneh Boroumand, Evan Smith, Pedro Figueiredo, Robert E. Peale
Department of Physics, University of Central Florida, Orlando, FL, USA 32816

ABSTRACT

We report a method for stress measurement and analysis in silicon oxide thin films using optical interference. Effects of design and fabrication on stress have been studied by fabricating submicron-thick slabs of oxide anchored at one end and extending over a reflective surface. Optical interference occurs between reflections from the surface and the oxide slab, giving rise to light and dark fringes that may be imaged with a microscope. Analysis of the interference pattern at different wavelengths gives the radius of curvature and means of stress mapping. The accuracy exceeds non-interferometric profilometry using optical or confocal microscopes, and it can be more quantitative than scanning electron microscopy. This nondestructive profilometry method can aid the stress optimization of silicon oxide or other transparent thin films to achieve specific mechanical characteristics in MEMS devices.

INTRODUCTION

We report an optical interference method to measure stress in a silicon dioxide thin film. This method is based on observation of Fizeau fringes [1] that are caused by interference of reflected light between a curved semi-reflective silicon dioxide thin film and a flat reflective surface beneath it. Fizeau interferometry is widely used to compare the shape of an optical surface on a mirror or lens to a reference surface of known shape [2]. The two surfaces are separated by a narrow gap, and interference fringes in reflected monochromatic light indicate spatial variations of the gap. Among other applications are thickness measurement of thin films, strain measurement of fiber optics, residual wedge measurement for optical flats and characterization of organic light emitting devices [3-6].

Stress is important to free-standing thin films in MEMS due to the deformations it induces, intended or otherwise. Intrinsic stresses, which depend on deposition conditions, are difficult to predict. Usual methods to measure stress in a thin film require measurement of the radius of curvature of a large substrate (e.g. a wafer) on which the film has been deposited and to which it is firmly attached [7]. This can be either done by a contact profiler, which can damage soft and suspended features, or by noncontact profilers, which can be expensive and slow.

We are interested in controlling the stress and deformation in free standing MEMS cantilevers, which consist of a 500 nm thick oxide topped with 30 nm of Cr/Au above a gold surface plate (Figure 1). Observed Fizeau fringes allow observation of height and curvature, as shown in Figure 1 (right). Cantilever motion and curling appear as a change in the fringe pattern. These cantilevers tend to curl upward after the metal deposition and release due to the different thermal expansion coefficients of metal and oxide. The curvature depends on the oxide deposition recipe, where different methods give different intrinsic stress, and on the temperature of the sample during deposition.

The 500 nm low stress silicon dioxide was deposited by Trion Orion II PECVD system using TEOS gas on top of a polyimide sacrificial layer. The cantilever legs are attached to the substrate via small square anchors. Photolithography, e-beam evaporation of 5 nm Cr and 25 nm

Au, and lift-off creates a dry etch mask for both oxide (CF_4 etch) and polyimide layers (O_2 etch). After release we obtain the structure imaged by scanning electron microscopy (SEM) in Figure 1 (left). Figure 1 (right) shows an optical microscope image of a different cantilever in monochromatic light showing Fizeau fringes in the form of rings. To characterize the results of experiments, we developed the optical interferometric method described in this paper.

Figure 1 (left) - SEM image of 500 nm thick SiO_2 cantilever with semitransparent Au surface coating suspended over a Au surface plate. The cantilever is curled up due to intrinsic oxide stress and to differential thermal expansion, since the sample temperature exceeded ambient during the Au deposition. (Right) Optical microscope image in monochromatic light, showing Fizeau fringes.

EXPERIMENTAL DETAILS

Simplified cantilevers with a range of widths and lengths were fabricated. Figure 2 (left) presents a schematic of the processing steps. We spin-coated a Si wafer with 1.2 um ProLift-100 as polyimide sacrificial layer, and then 600 nm PMMA (495 A) was spin-coated on top. The desired pattern was exposed by electron-beam to define 10 micron square anchors. The PMMA was developed by MIBK:IPA 1:3 solution and ProLift was etched 15 s with TMAH based developer MF319 following by 75 sec dry etch in plasma enhanced etcher with O_2 gas. Longer wet etch undercuts the ProLift. PMMA was stripped in acetone, and 500 nm TEOS-based silicon oxide was deposited on the ProLift using the Trion PECVD system. The recipe was optimized to achieve high step coverage to strengthen the anchor neck points. The cantilever etch mask was produced by another PMMA spin, e-beam exposure, and MIBK:IPA development, followed by 42 nm sputtered Au and lift off. The Au serves as the reactive ion etch (RIE) mask for etching the oxide in CF_4 gas. The last steps to release the cantilevers are 2 min anisotropic etch of Prolift in RIE system using O_2 gas mixed with 6% CF_4 (300 W, 100 mTorr) and 10 min isotropic etch (300 W, 300 mTorr) while the sample is tilted 45 deg [8]. Figure 2 (right) presents an optical microscope image of the resulting cantilever array.

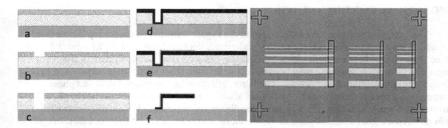

Figure 2 (left) Fabrication steps of cantilevers: a. Spin Prolift 100-2 as sacrificial layer, then PMMA; b. Pattern PMMA using e-beam writer and development in MIBK:IPA solution; c. Transfer pattern to sacrificial layer using combination of wet and dry etch; d. PECVD SiO$_2$; e. Patterned Au lift-off to achieve oxide etch mask; f. Etch oxide in RIE, then release in O$_2$ plasma RIE. (right) Optical microscope image of cantilever arms with length 55, 120, and 250 micron and with width 1, 5, 10, and 25 micron. The narrowest arms are invisible in this image. All arms are anchored at one end. The image was collected before etching the oxide and release from the sacrificial layer.

Fizeau fringes were recorded with a microscope equipped with a digital CCD camera. Images were analyzed in Labview to obtain line intensity profiles. To enhance fringe visibility and allow quantitative analysis, either a monochromatic laser or narrow band-pass filtered white light were used for illumination (Figure 3).

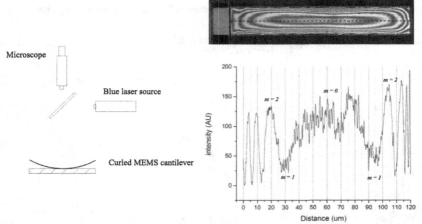

Figure 3 (left) – Schematic of set-up for observing Fizeau fringes. (right) Microscope image of cantilever with results of indicated intensity line-scan.

THEORETICAL CONSIDERATIONS

The optical boundaries that reflect light in the Figure 2 structures are the top Au surface, the Au/SiO$_2$ boundary, the SiO$_2$-air boundary underneath the cantilever, and the Si substrate surface. Interference between reflections from the top three parallel surfaces gives no fringes. Fringes due to interference come only between reflected light from Si substrate surface and light reflected from the cantilever as a whole. The latter reflection has some amplitude and phase whose exact values are unimportant. Amplitude affects fringe visibility while a shift in phase is equivalent to a uniform height offset between cantilever and substrate. We are only interested in height differences from different parts of the structure, which for adjacent light and dark fringes is just half a wavelength $\lambda/2$.

The profile of small deformations may be considered as the arc of a circle of radius R, as shown in the Figure 4 schematic. Stoney's formula [9] relates radius of curvature in a double layer structure to the stress in the film as

$$\sigma^{(f)} = \frac{E_{SiO2}}{6(1-v_{SiO2})} \times \frac{h_{SiO2}^2}{Rh_{Au}} \tag{1}$$

where E and v are Young's modulus and Poisson's ratio respectively and h is the layer thickness. This formula is valid when $h_{Au} \ll h_{SiO2} \ll R$. The height differences d_m above the minimum at the position of the mth ring is $m\lambda/2$, where m is an integer that is incremented with each new light or dark ring counting from the central spot (m = 0). See Figure 3 for an example of how the rings are numbered. Across a cantilever d_m generally amounts to only several half wavelengths, i.e. no more than a few microns, while ring radii r_m are on the scale of 10s of microns, according to Figure 3. In this limit of $d_m \ll r_m$, $R \approx \frac{r^2}{2d}$ according to Figure 4 so that Eq. 1 becomes

$$\sigma^{(f)}(r) = \frac{E_{SiO2}}{6(1-v_{SiO2})} \times \frac{h_{SiO2}^2}{h_{Au}} \times \frac{m\lambda}{r_m^2} \tag{2}$$

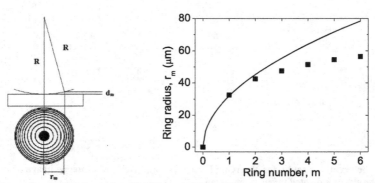

Figure 4 – (left) Schematic with air gap d_m, ring radius r_m, and radius of curvature R. (right) ring radius r_m vs. ring number m for Figure 3 profile (symbols) and function $32\sqrt{m}$ (line).

RESULTS AND DISCUSSION

The assumption that the deformation along a particular direction is the arc of a circle means the stress has the same value at every point along that direction. In other words, uniform σ has no m dependence, which requires r_m to increase as the square root of m according to Eq. 2. For the example of Figure 4 (right), the experimental r_m values rise more slowly than $\sqrt{m}$, which implies that stress is higher near the edges. In other words, closer ring spacing means more curling, which implies higher stress. For other directions the stress might be lower at the edges. That stress is not uniform is supported by Figure 5, where the Fizeau rings even lack the same symmetry as the cantilever, one corner being strongly curled.

Figure 5a presents an SEM image of one of the fabricated arms. Curling of the lower right corner is obvious, but it is clearly impossible from this image to quantify the deformation. Figure 5b presents the Fizeau fringes for the same arm at 408 nm wavelength. To obtain a map of stress over the surface, a radial mesh was drawn from the center dark fringe to the boundaries and the position of each dark and bright fringe was determined along each line. Figure 5c is the resulting contour plot of the d_m in units of μm. Figure 5d gives the stress distribution over the surface calculated according to Eq. 2, where the darkest shade indicates 111 MPa and the lightest 753 MPa The stress is highest along the short direction and at the curled corner and lowest in the long direction. During release a tilted sample helps RIE removal of hard baked Prolift. This may explain the asymmetry of the deformation [10].

In summary, we have presented a method of measuring topography, stress (and motion) of free standing transparent films with high spatial resolution and without special instrumentation.

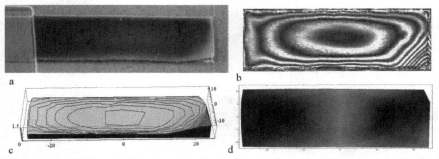

Figure 5 – a. SEM image of an arm after release b. Image taken with 408 nm wavelength source revealing fringe pattern. c. Contour plot. d. Calculated stress map on the surface of cantilever, bright areas shows higher stress values.

ACKNOWLEDGMENTS

This work was supported in part by a contract EMX International LLC, and by a grant from the Florida High Technology Corridor (I-4) program. Travel to attend this meeting was supported by UCF Graduate Studies and the UCF Student Government Association.

REFERENCES

1. A. Reserbat-Plantey, L. Marty, O. Arcizet, N. Bendiab, V. Bouchiat, Nature Nanotechnology 7, 151 (2012).
2. NASA preferred reliability practices. GUIDELINE NO. GT-TE-2404, NASA Technical Memorandum 4322A, NASA Reliability Preferred Practices for Design and Test, (NASA Office of Safety and Mission Assurance, Washington, 1999).
3. J. G. Gottling and W. S. Nicol, JOSA 56, 1227 (1966).
4. E. Li, G. Peng, X. Ding, Appl. Phys. Lett. 92, 101117 (2008).
5. S. Chatterjee, Opt. Eng. 42, 3235 (2003).
6. C. Tsai, K. Tien, M. Chen, K. Chang, M. Lin, H. Cheng, Y. Lin, H. Chang, H. Lin, C. Lin, C. Wu, Organic Electronics 11, 439 (2010).
7. T. M. Adams and R. A. Layton, *Introductory MEMS: Fabrication and Applications* (Springer, Berlin, 2010).
8. J. Boroumand Azad, I. Rezadad, J. Nath, E. Smith, R. E. Peale, Proc. SPIE 8682-80 (2013).
9. G. G. Stoney, Proc. R. Soc. London, Ser. A 82,172 (1909).
10. A. A. Volinsky, G. Kravchenko, P. Waters, J. D. Reddy, C. Locke, C. Frewin, S. E. Saddow in *Residual Stress in CVD-grown 3C-SiC Films on Si Substrates,* edited by M. Law, B. J. Pawlak, M. L. Pelaz, K. Suguro (Mater. Res. Soc. Symp. Proc. 1069, 1065-D03-05, 2008)

Mater. Res. Soc. Symp. Proc. Vol. 1536 © 2013 Materials Research Society
DOI: 10.1557/opl.2013.924

Effect of RF or VHF Plasma on Nanocrystalline Silicon Thin Film Structure: Insight from OES and Langmuir Probe Measurements

Lala Zhu[1,2], Ujjwal K Das[1] , Steven S Hegedus[1], Robert W Birkmire[1,2]
1. Institute of Energy Conversion, University of Delaware, Newark, Delaware, USA 19716;
2. Department of Physics and Astronomy, University of Delaware, Newark, Delaware, USA 19716.

ABSTRACT

Optical emission spectroscopy (OES) and Langmuir Probe were used to characterize RF and VHF plasma properties under conditions leading to nanocrystalline silicon film deposition. Films deposited by RF plasma at low pressure (3 Torr), even with high crystalline volume fraction, show weak X-ray diffraction signals, suggesting small grain size, while RF films at higher pressure (8 Torr) and VHF films at both high and low pressure have larger grain sizes. The preferential growth orientation is controlled by the H_2/SiH_4 ratio with RF plasma, while the film deposited by VHF shows primarily (220) orientation independent of H-dilution ratio. Langmuir Probe measurements indicate that the high energy electron population is reduced by increasing pressure from 3 Torr to 8 Torr in RF plasma. Compared with RF plasma, the VHF plasma shows higher electron density and sheath potential, but lower average electron energy, which may be responsible for the larger grain size and crystal orientation. The growth rate and crystalline volume fraction of the film is correlated with OES intensity ratio of SiH* and H_α/SiH* for both RF and VHF plasmas.

INTRODUCTION

Nanocrystalline silicon (nc-Si) is an attractive material for thin film photovoltaic application due to its lower band gap and better stability compared with amorphous silicon (a-Si). It also performs better in thin film transistors (TFT) due to higher mobility. nc-Si films are commonly deposited by plasma-enhanced chemical vapor deposition (PECVD) with silane gas (SiH_4) highly diluted with hydrogen (H_2). Due to the indirect band gap of nc-Si, a much larger thickness is required for light absorption, so growth rate is critical for high throughput in industry processing. The best intrinsic nc-Si film for solar cell application is deposited at the amorphous/nanocrystalline transition region [1]. It has been reported that better nc-Si cell performance is obtained with (220) oriented films and the grain size around 25 nm [2]. Optical emission spectroscopy (OES) provides the intensity of emission species. The plasma's electrical properties, such as electron density, average electron energy, electron energy distribution [3] and sheath potential, were measured by Langmuir Probe.

In this study, nc-Si films were deposited by RF (13.56 MHz) and VHF (40.68 MHz) plasma. The combination of OES and Langmuir Probe measurements were used to evaluate the plasma properties, which can be correlated to film properties of growth rate, crystalline volume fraction, grain size and crystal orientation.

EXPERIMENT

nc-Si:H films were deposited in a PECVD system either using capacitively coupled RF or VHF plasma. SiH_4 and H_2 were delivered into the reactor through a gas showerhead, where the area of the powered electrode is 1100 cm^2 and the discharge gap is 1.4 cm. Films were deposited at different excitation frequency, plasma power, pressure and H_2 dilution. The H_2 dilution was varied by changing the SiH_4 flow rate at a fixed H_2 flow rate. The films were deposited on Corning 1737 glass at a temperature of 200^0C with a thickness of $\sim 1 \mu m$. It is known that the nanocrystalline volume fraction typically increases with film thickness at a given deposition condition[4]. However, the films deposited on to a seed layer were found to minimize such thickness dependence of crystalline fraction [5]. In this paper, prior to each nc-Si film deposition, a short (2 mins) H_2 plasma treatment followed by a thin (10 nm) seed layer was deposited to avoid film peeling and improve structural homogeneity respectively.

The growth rate (GR) was calculated by the film thickness estimated from a Dektak stylus profiler. Raman spectroscopy was performed with 532 nm laser excitation and the crystalline volume fractions (X_c) were calculated from the three peaks deconvolution method [5]: $X_c=(I_{520}+I_{510})/(I_{520}+I_{510}+I_{480})$. The films were measured by X-ray Diffraction (XRD) to investigate the Si (111), (220) and (311) peaks. The growth orientation was estimated by peak intensity ratios and grain size (GS) was calculated by the full width at half maximum (FWHM) of different peaks using the Scherrer equation.

OES spectra were recorded from the plasma emission through a quartz window. The Si* (288nm), SiH* (414nm), H_α (656nm) and H_β (486nm) peak intensities were identified by OES. A Langmuir Probe with a single tungsten wire, 0.15 mm diameter and 10 mm length, was inserted into the plasma to measure the current with a fast voltage sweep from -50 V to +50 V. The probe was cleaned by electron current heating at +70 V bias in Ar plasma. Sheath potential was obtained by subtraction of plasma potential (V_p) and floating potential (V_f). The electron energy distribution function (EEDF), f(E), was obtained from the Druyvesteyn expression[3,6]. Electron density (n_e) and mean electron energy ($<E_e>$) were calculated from f(E):

$$f(E) = \frac{4}{e^2 A} \sqrt{\left(\frac{-mV}{2e}\right)} \frac{d^2 I}{dV^2} \; ; \; n_e = \int f(E)dE \; ; \; <E_e> = \frac{1}{n_e} \int E * f(E)dE.$$

where V and I are voltage and current obtained from the Langmuir Probe, e and m are electron charge and mass, A is the probe area, and the electron energy is $E=-e(V-V_p)$.

RESULTS AND DISCUSSION

nc-Si films structure and plasma properties with RF discharge

Si thin films were deposited at 3 Torr, RF-power of 250W with a constant H_2 flow of 450 sccm. The SiH_4 flow was increased in 1 sccm steps from 7 sccm to 14 sccm. GR increased from 1.7 A/s to 3.2 A/s, while X_c decreased from 68% (nc-Si) to 0% (a-Si). SiH* and H_α emission intensities were collected simultaneously to correlate with the film properties. Figure 1 (a) presents the variations of SiH* intensity and the growth rate at different H_2/SiH_4 ratios. Higher SiH_4 partial pressure increases the generation rate of silicon species resulting in a higher SiH* intensity and GR. Figure 1 (b) presents the ratio of $H_\alpha/SiH*$ and X_c of the deposited films. The $H_\alpha/SiH*$ decrease by increasing SiH_4 flow is mainly due to higher Si species generation and H elimination by the reaction $H+SiH_4 \rightarrow SiH_3+H_2$. The lower atomic H flux to the growing surface results in the lower X_c. The nc-Si to a-Si transition occurs when SiH_4 flow increases from 13 sccm to 14 sccm, with a $H_\alpha/SiH*$ threshold value between 3.8 and 4.3 in this series.

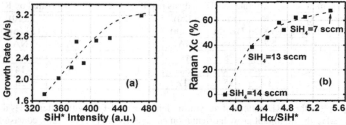

Figure 1: OES result by RF (a) SiH* intensity and growth rate, (b) Hα/SiH* and Raman Xc
(Lines are drawn to guide the eye)

A further series of nc-Si films were deposited by RF to explore the process conditions and final film structure. Figure 2 (a) shows the XRD patterns of films deposited at different pressures, from 3 Torr to 8 Torr, with $H_2/SiH_4=300/8.5$ and RF power 515 W. Table 1 lists the analysis results from XRD, as well as GR and X_c. At 3 Torr, there is no significant XRD peaks observed, indicating small crystal grain size. By increasing the pressure to 5 Torr, grain size calculated from either the (111) or (220) peaks increases. The integrated intensity ratio $I_{(220)}/I_{(111)}=0.67$ indicates the film is almost randomly oriented. Grain size continues to increase with higher deposition pressures. At 8 Torr, film deposition is near the amorphous/crystalline phase transition region with X_c around 50%. A grain size 24 nm is observed calculated from the (220) peak and the film is strongly (220) oriented. Figure 2(b) presents normalized Raman spectra of the films deposited by RF at 515 W. A strong crystalline phase peak (~520 cm^{-1}) is confirmed for the film deposited at 3 Torr, even though no XRD peaks were detected. By increasing the pressure from 3 Torr to 8 Torr, the Raman peak shifts from 517 cm^{-1} to 519 cm^{-1} consistent with larger grain size. The FWHM of crystalline peak (~520 cm^{-1}) decreases from 15 cm^{-1} to 10 cm^{-1}. The larger FWHM at 3 Torr may also indicate a wider grain size distribution [7] in the film.

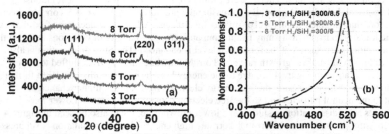

Figure 2: (a) XRD pattern of nc-Si film by RF 515W at different pressure, (b) Raman spectra of nc-Si film at 3 Torr and 8 Torr by RF 515W

Table 1: Raman Xc, growth rate, XRD analysis results for nc-Si film deposited by RF

Pressure (Torr)	X_c (%)	GR (A/s)	(111) Height	(220) Height	$I_{(220)}/I_{(111)}$	(111) GS (nm)	(220) GS (nm)
3	67	2.7	NA	NA	NA	NA	NA
5	79	4.1	98	49	0.67	13	10
6	82	3.7	100	84	1.43	16	14
8	52	3	35	312	7.9	16	24

Figure 3 presents the XRD pattern of nc-Si film deposited at 8 Torr with different H_2 dilutions. By reducing H_2/SiH_4, the dominant growth orientation changes from (111) to (220), consistent with results reported by others [2].

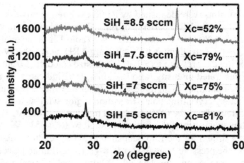

Figure 3: XRD pattern of nc-Si films different dilution at 8 Torr by RF 515W

Langmuir probe was used to measure the EEDF at different pressures in RF plasma (Fig.4(a)), and the electron density and mean energy were estimated. Figure 4 (b) presents the relationship between pressure and electron density. By increasing the pressure from 3 Torr to 5 Torr, the electron density is more than doubled from 1.4×10^{15} m^{-3} to 3.3×10^{15} m^{-3}, while $<E_e>$ decreases from 8.0 eV to 3.8 eV. The higher electron density may result from an increasing number of collisions between the electrons and molecules, such that the ionization level is enhanced. The SiH_4 dissociation is also increased leading to the highest growth rate at 5 Torr. Meanwhile, the EEDF is shifted to lower energy values, as shown in Figure 4 (a), due to the energy loss during the collisions. With further increases in the deposition pressure, the low energy electrons have a much higher chance of attachment to the neutral particles and so the free electron density is decreased. Lower free electron density reduces SiH_4 dissociation, resulting in a lower growth rate. It can also be seen in Figure 4 (a) that, by increasing pressure from 3 Torr to 8 Torr, the high energy electron population is suppressed. The sheath potential, as shown in Figure 4(d), decreases from 24 V at 3 Torr to 14 V at 8 Torr, indicating less ion bombardment energy at high pressure. It has been reported that the grain size could be reduced by the excess ion bombardment at a high power regime [8]. The larger grain size at higher pressure by RF may result from the combination of low energy electron-molecule impacts and less ion bombardment damage.

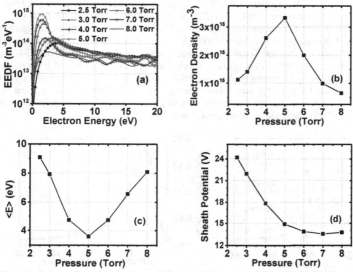

Figure 4: Langmuir probe results at different pressure by RF (a) EEDF, (b) Electron density,
(c) Mean electron energy, (d) Sheath potential

nc-Si films structure and plasma properties with VHF discharge

A series of nc-Si films were deposited by VHF with fixed deposition parameters, 3 Torr, 250W, H_2 =300 sccm, while SiH_4 flow was varied from 8.5 sccm to 25.5 sccm. The results shown in Figures 5(a) and 5(b) indicate the growth rate and crystalline volume fraction strongly depend on the SiH* intensity and H_α/SiH*, similar to that observed from the RF results.

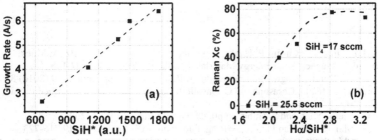

Figure 5: OES result by VHF (a) SiH* intensity and growth rate; (b) H_α/SiH* and Raman X_c
(Lines are drawn to guide the eye)

Figure 6 presents the XRD pattern of the films in this series at different dilution. The grain size from the VHF films, calculated from FWHM of (220) peak, is ~27 nm, while for RF films no significant peak was observed at comparably high X_c ~ 75%. Unlike the trend shown in Figure 3(b) for RF films, films deposited by VHF are all (220) preferential orientation (Figure 6), independent of the H_2/SiH_4 ratio. Figure 7 shows the EEDF (normalized to unit integral area) for RF and VHF at 3 Torr, 250W. The $<E_e>$ is 5.6 eV for VHF (large grain size) and 9.2 eV for RF (small grain size). The sheath potential, measured by Langmuir probe, for VHF is around 36 V for all the conditions in this series, ~10 V higher than RF at 3 Torr, which indicates the VHF has higher ion bombardment energy than RF in this study. However, the ion flux on the growth surface should be lower at VHF due to low $<E_e>$, which results in lower ion bombardment damage.

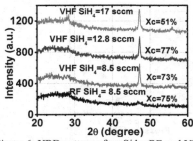

Figure 6: XRD pattern of nc-Si by RF and VHF   Figure 7: EEDF of RF and VHF plasma

CONCLUSIONS

nc-Si films were deposited by RF and VHF plasma while varying the H_2-dilution and pressure. The plasma properties were investigated by OES and Langmuir Probe. For both RF and VHF films, OES SiH* intensity and H_α/SiH* is correlated with the growth rate and crystalline volume fraction. The preferential growth orientation with RF depends on H_2/SiH_4 ratio, while films deposited by VHF are all strongly (220) oriented independent of dilution. Larger grain size could be obtained by either plasma with low mean electron energy. There is no strong correlation between grain size and sheath potential.

REFERENCES

1. O. Vetterl, F. Finger, R. Carius, P. Hapke, L. Houben, O. Kluth, A. Lambertz, A. Muck, B. Rech and H. Wagner, Solar Energy Materials & Solar Cells 62 (2000) 97-108
2. K.Saito and M. Kondo, Progress in Photovoltaics: Research and Applications 19 (2010) 858–864
3. T. Moiseev, G. Isella, D. Chrastina and C. Cavallotti, Journal of Physics D: Applied Physics 42 (2009) 225202
4. B. Yan, G. Yue, J. Yang and S. Guha, Applied Physics Letters 85,1955 (2004)
5. C. Ross, Y. Mai, R.Carius and F. Finger Prog. Photovolt: Res. Appl. 2011; 19:715-723
6. Druyvesteyn M J 1930 Z. Phys. 64 781–98
7. W. Ke, X. Feng and Y. Huang, Journal of Applied Physics 109 (2011) 083526
8. M. Kondo, M. Fukawa, L. Guo and A. Matsuda, Journal of Non-Crystalline Solids 266-269 (2000) 84-89

Defects and Transport

Mater. Res. Soc. Symp. Proc. Vol. 1536 © 2013 Materials Research Society
DOI: 10.1557/opl.2013.786

Defect Densities and Carrier Lifetimes in Oxygen doped Nanocrystalline Si

Shantan Kajjam[1], Siva Konduri[1], Max Noack[2] , G. Shamshimov[3], N. Ussembayev[3] and Vikram Dalal[1]

[1]Department of Electrical & Computer Engineering, Iowa State University, Ames IA 50014
[2]Microelectronics Research Center, Iowa State University, Ames IA 50014
[3]Nazarbayev University, Astana, Kazakhstan

ABSTRACT

We report on the measurement of defect densities and minority carrier lifetimes in nanocrystalline Si samples contaminated with controlled amounts of oxygen. Two different measurement techniques, a capacitance-frequency (CF) and high temperature capacitance-voltage techniques were used. CF measurement is found to yield noisy defect profiles that could lead to inconclusive results. In this paper, we show an innovative technique to remove the noise and obtain clean data using wavelet transforms. This helps us discover that oxygen is creating both shallow and deep/midgap defect states in lieu with crystalline silicon. Minority carrier lifetime measured using reverse recovery techniques shows excellent inverse correlation between deep defects and minority carrier lifetimes through which hole capture cross section can be evaluated.

INTRODUCTION

Nanocrystalline Silicon (nc-Si) is an important material for solar cells and thin film transistors [1]. Its fabrication using plasma enhanced chemical vapor deposition (PECVD) is highly popular. Due to its method of fabrication, oxygen becomes a non-ignorable impurity and hence is important to study its degrading effects on device parameters. T. Kilper et al [2] have discussed about reduction in quantum efficiency due to presence of oxygen in the material that was introduced as a chamber leak. P. Hugger et al [3] showed variation in defect density versus oxygen content in the material. There is no previous study to indicate how the profile of defects varies within the band gap of nc-Si.

We introduce oxygen systematically into the material and study the variation in defect densities, defect profiles and lifetimes. We also discuss the problem associated with measurement of defect profiles and provide a solution by using wavelet transform techniques and Matlab [4].

EXPERIMENT

Nc-Si solar cells have been fabricated using very high frequency PECVD (45MHz) on stainless steel substrates. An n-i-p stack is made on the substrate with a-Si n+ and a thin nc-Si p+ at the top. The i-layer is made using hydrogen profile technique with silane and hydrogen at high dilution to attain an optimum crystallinity and a thickness of ~1.3-1.5μm. Controlled oxygen doping is done by using a 108ppm O_2/He cylinder. Device contacts are ITO and sample is mesa-etched to reduce capacitance fringing. By covering ITO dots with wax and reacting the device with HF and HNO_3, we can remove the non-waxed area of the device, thus leaving only the dotted lined structure below the ITO contact in figure 1, and thereby reduce the influence of fringing fields on capacitance.

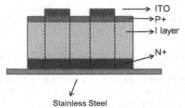

Figure 1. Schematic of the device

Measurement Techniques

Capacitance Vs voltage (CV) is a proven technique for measuring traps at deep & shallow levels. By plotting $(A/C)^2$ versus voltage (A is the area of device) and extracting the slope, the defects in the region can be calculated [5]. The number of defects responding can be controlled by applied frequency & sample temperature.

Another important technique is to use capacitance – frequency (CF) analysis which allow for an in-depth study of defect profiles [6]. The deeper in energy a trap is, the longer it takes for it to emit its electron and the lower the frequency needed to measure it. By taking a differential in capacitance over frequency, the capacitance contribution of trap states at an energy level can be calculated using equations 1 & 2.

$$N_t(E_\omega) = -\frac{U_d}{qw}\frac{dC}{d\omega}\frac{\omega}{kT} \qquad (1)$$

$$E_\omega = kT \ln\left(\frac{\upsilon_o}{\omega}\right) \qquad (2)$$

In these equations, C is the capacitance per unit area, ω is the angular frequency, E_ω is the energy of the trap below conduction band, U_d is the built-in voltage at -1V reverse bias, υ_o is the attempt to escape frequency, w is the depletion width, q is the charge of electron, T is the temperature and k is the Boltzmann constant.

Carrier lifetime was measured using the Reverse Recovery Transient method [7]. In this method a diode in forward bias is switched abruptly to reverse bias. The stored minority carriers in the n layer of the p-n junction do not fall to zero abruptly and the current takes time to reduce to reverse saturation current which is a direct indicator of lifetime. Initial measurements of lifetime in nc-Si have been done by S. Saripalli[8].

RESULTS

CV has been measured at 300 kHz, 23^{0}C for dopant density and at 20 Hz, 100^{0}C for total defect density (Figure 2). The variation of oxygen is indicated by the flow ratio of oxygen to silane in ppm. A steady increase in both the shallow defects and the deep defects has been observed.

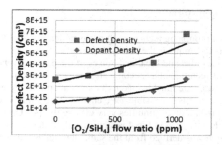

Figure 2: Dopant & Deep Defects variation with oxygen flow.

On a standard non-oxygen doped sample, CF measurement has been done between 20Hz – 200kHz at -1V reverse bias and room temperature. Upon using equations (1) & (2), we can evaluate the trap profile below the conduction band. Figure 3 shows the experimental CF data and Fig. 4 shows the evaluated trap profiles vs. energy obtained from the data in Fig. 3. As can be seen, there is a significant amount of noise in the data of Fig. 4 which is mainly arising due to the presence of a differential component in equation 1.

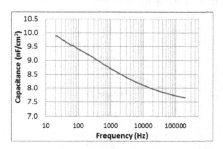

Figure 3: Capacitance - Frequency data for nc-Si sample

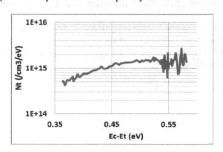

Figure 4: Trap density vs. trap energy evaluated from data of Fig. 3.

Denoising using Wavelet Transform

Wavelet analysis is a promising technique to perform denoising of the data in Figure 3. The major advantage of this technique is the ability to perform localized signal analysis. It is capable of revealing discontinuities and localized noises of the data that other techniques like fourier transform miss. This technique has become an indispensable solution for noise analysis [4]. For the defect profile data, care must be taken that the data points are equidistantly spaced along the x-axis so that this analysis is possible. We choose Daubechies wavelet for our denoising because it picks up data discontinuities very efficiently. [4]. Upon denoising using wavelets, the noise in trap density seen in Fig.4 is significantly reduced, as seen in the 0 ppm oxygen curve in Fig. 5, which was derived from the same data set as the curve in Fig. 4.

Variation of Defect Profile with Oxygen

The defect profiles are denoised and plotted for varying oxygen flow in Figure 5. It can be seen that there is an increase in the energy range being measured, including at 0.38eV & 0.51eV. These are the levels that correspond to oxygen in crystalline Si. [9]. The broad range of energy over which the defects exist suggests that the defects in the a-Si tissue surrounding the grain are also increasing with oxygen doping.

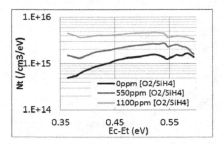

Figure 5: Defect Profiles with variation in Oxygen flow.

In Figure 6, we plot the lifetime vs. inverse of defect density, yielding a straight line going through the origin. This curve indicates that the trap-controlled recombination mechanism is dominating in this material. From the slope of the curve, we can calculate the capture cross-section of the traps to be $\sim 3 \times 10^{-17}$ cm^2.

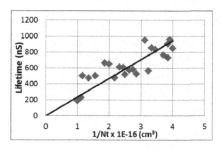

Figure 6. Plot of minority carrier lifetime vs. inverse of defect density

CONCLUSIONS

In conclusion, we show that oxygen is increasing the defect density in nanocrystalline Si. The defects appear to be in both within the crystalline grain, and at the grain boundaries. The minority carrier lifetime is inversely related to the defect density.

ACKNOWLEDGMENTS

We thank the funding agencies – NSF and Center for Energy Research, Nazarbayev University, Kazakhstan for jointly funding this project. We are grateful to all the graduate and undergraduate students at Microelectronics Research Center for their help.

REFERENCES

1. Shah, J. Meier, E. Vallat-Sauvain, N. Wyrsch, U. Kroll, C. Droz, U. Graf, Solar Energy Materials and Solar Cells 78 (2003) 469-491.
2. T. Kilper, W. Beyer, G. Brauer, T. Bronger, R. Carius, M. van den Donker, D. Hrunski, A. Lambertz, T. Merdzhanova, A. Muck, Journal of Applied Physics 105 (2009) 074509-074509-074510.
3. P.G. Hugger, J.D. Cohen, B. Yan, G. Yue, J. Yang, S. Guha, Applied Physics Letters 97 (2010) 252103.
4. Misiti, Michel, et al. "Wavelet toolbox." The MathWorks Inc., Natick, *MA* (1996).
5. L. C. Kimerling, J. Appl. Phys. **45**, 1839 (1974)
6. T. Walter, R. Herberholz, C. Muller, H. Schock, Journal of Applied Physics 80 (1996) 4411-4420.
7. B. Lax, S. Neustadter, Journal of Applied Physics 25 (1954) 1148-1154.
8. S. Saripalli, V. Dalal, Journal of Non-Crystalline Solids 354.19 (2008): 2426-2429.
9. Sze, Simon M., and Kwok K. Ng. "Physics of Semiconductor Devices". (Wiley-Interscience, 2006).

Mater. Res. Soc. Symp. Proc. Vol. 1536 © 2013 Materials Research Society
DOI: 10.1557/opl.2013.746

Microstructure Characterization of Amorphous Silicon Films by Effusion Measurements of Implanted Helium

W. Beyer[1, 2, 3], W. Hilgers[2], D. Lennartz[2], F.C. Maier[3], N.H. Nickel[1], F. Pennartz[2], P. Prunici[3]

[1]Institut für Silizium-Photovoltaik, Helmholtz-Zentrum Berlin für Materialien und Energie, Kekuléstrasse 5, D-12489 Berlin, Germany
[2]IEK5-Photovoltaik, Forschungszentrum Jülich GmbH, D-52425 Jülich, Germany
[3]Malibu GmbH & Co.KG, Böttcherstrasse 7, D-33609, Bielefeld, Germany

ABSTRACT

An important property of thin film silicon and related materials is the microstructure which may involve the presence of interconnected and isolated voids. We report on effusion measurements of implanted helium (He) to detect such voids. Several series of hydrogenated and unhydrogenated amorphous silicon films prepared by the methods of plasma deposition, hot wire deposition and vacuum evaporation were investigated. The results show common features like a He effusion peak at low temperatures attributed to He out-diffusion through a compact material or through interconnected voids, and a He effusion peak at high temperatures attributed to He trapped in isolated voids. While undoped plasma-grown device-grade hydrogenated amorphous silicon (a-Si:H) films show a rather low concentration of such isolated voids, its concentration can be rather high in doped a-Si:H, in unhydrogenated evaporated material and others.

INTRODUCTION

Void-related microstructure is known to be an important property of thin film silicon materials. Voids, i.e. volume parts which are empty or have a strongly reduced mass density may cause electronic defect states like dangling bonds or strained Si-Si bonds as they act as internal surfaces. Effusion measurements which detect gases leaving a given material on heating can often distinguish between compact material which may contain isolated voids and material with interconnected voids [1]. Interconnected voids, i.e. channels of reduced mass density throughout the material may provide rapid out-diffusion paths for hydrogen molecules as well as in-diffusion paths for oxygen or water molecules. Reduced electronic quality of amorphous silicon (a-Si) and related alloys has often been associated with the presence of clustered hydrogen and hydrogen-related interconnected voids [2]. Isolated voids, on the other hand, may trap gas atoms or molecules and high pressure molecular hydrogen in such voids (of 10-20 Å diameter) has been reported [3]. Here we use effusion of implanted helium to detect and characterize void-related microstructure. He effusion curves are expected to give information on material imperfections since He does not react with the silicon host material [4]. Low temperature He effusion has been associated with interconnected voids [4]. A high temperature He effusion stage has been attributed to the presence of isolated voids and a parameter F* was defined to estimate the isolated void concentration [5]. We report on effusion measurements of hydrogen and implanted helium for series films of hydrogenated amorphous silicon (a-Si:H) prepared by the methods of plasma deposition (PD) and hot wire (HW) deposition, and of (nominally) unhydrogenated amorphous silicon deposited by evaporation in vacuum (EV).

EXPERIMENT

The investigated films were typically 1 μm in thickness, deposited on crystalline Si wafer material. Helium was implanted at the energy of 40 keV and at the low doses 3×10^{15} and 10^{16} cm^{-2}. Implantation energy of 40 keV corresponds to a mean helium depth of about 0.35 μm. Helium (and hydrogen) effusion measurements were performed as reported elsewhere [1] using a heating rate of 20°C per minute. By integrating over the low temperature (T< 600°C) and high temperature (T > 600°C) helium effusion, the ratio F* was determined [5]. The a-Si materials investigated involved (undoped) material prepared at Berlin by e-beam vacuum evaporation, (undoped) a-Si:H deposited at Jülich by HW deposition [4, 6] and a-Si:H (undoped and doped by diborane and phosphine flows) deposited at Jülich by conventional rf plasma using undiluted silane. Substrate temperature T_S ranged between 25 and 500°C.

RESULTS AND DISCUSSION

In Figs.1a-e, 2a-d, and 3a-e effusion spectra of hydrogen and implanted helium are presented of undoped Si:H films fabricated by plasma deposition, hot wire deposition and vacuum evaporation, respectively, at a variety of substrate temperatures. Hydrogen effusion shows only for the plasma deposited film of $T_S = 50$°C two effusion peaks attributed to the out-diffusion of hydrogen molecules through a network of interconnected voids (near 400°C) and, following reconstruction of the material, to out-diffusion of atomic H (near 650°C). For all other samples, hydrogen shows a single more or less broad effusion maximum in the 600 – 700°C range indicating out-diffusion of H atoms. The temperature of this latter H effusion maximum depends usually on film thickness as this effusion process is rate-limited by H diffusion [1]. Note that the hydrogen content which for plasma and HW grown material decreases with rising T_S (typically 12 at. % at $T_S = 200$°C and 4 at. % at 400°C) is nearly constant for the evaporated

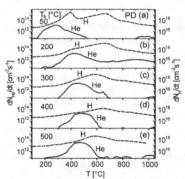 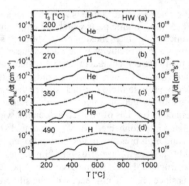

Figure 1. Effusion rate of H and implanted He (dose: 10^{16} cm^{-2}) for plasma deposited a-Si:H films using ((a)-(e)) various substrate temperatures T_S.

Figure 2. Effusion rate of H and implanted He (dose: 10^{16}cm^{-2}) for hot wire deposited a-Si:H [6] using ((a)-(d)) various substrate temperatures T_S.

material at H values near 0.5 at. %. For the plasma-deposited samples, He is seen to effuse in one or two stages. The lower temperature (LT) stage takes place below 600°C and can be characterized [4] by the temperature T_{He}^{LT} of maximum effusion rate. It has been attributed to He out-diffusion from compact material or through interconnected voids. For crystalline Si, we find for implanted He a peak near 370°C. Published data for He diffusion in crystalline Si [7] agree well with such a He effusion temperature if the relation $D/E_D = d^2 \beta/k\pi^2 T_{He}^{LT}$ (D: He diffusion coefficient, E_D: He diffusion energy, β: heating rate, k: Boltzmann constant) [1, 4] is used. $T_{He}^{LT} < 370°C$ is then attributed to the presence of interconnected voids, i.e. void channels where He can easily effuse out, while $T_{He}^{LT} > 370°C$ is associated with disordered dense Si regions. In compact a-Si, helium is likely to diffuse in a doorway diffusion process [1, 4, 8] and in such a process disorder presumably decreases the He diffusion coefficient and increases T_{He}^{LT}.

The higher temperature (HT) He effusion at T > 600°C (Fig. 1b, c) is in a temperature range where all implanted He should have effused out. In agreement with c-Si literature [9] we attribute this HT effusion process to the permeation of He trapped in isolated voids. These isolated voids are considered to be present from the deposition process. By measuring He effusion spectra as a function of He implantation dose it was found that He-implantation generated isolated voids appear in plasma deposited a-Si:H only at doses exceeding $10^{16} cm^{-2}$ [10]. However, some generation of isolated voids during the He or hydrogen effusion process (by He/ H_2 precipitation or by transformation of interconnected voids into isolated ones) cannot be excluded. It is thought that diffusing He (at T < 600°C) either reaches the film surface (or film-substrate interface) giving rise to the LT He effusion peak or gets trapped in isolated voids causing the HT peak. The results of Fig. 1 then show that isolated voids show up in our plasma deposited films only in a narrow substrate temperature range around $T_S = 200°C$. At $T_S > 300°C$, apparently the isolated void concentration is close to or below our detection limit.

Compared to the plasma deposited material, our (non-device grade) HW deposited films (Fig. 2) show more complicated He effusion spectra, although a general division into low-temperature processes below 600°C and high-temperature processes above 600°C is also possible. Samples with $T_S > 200°C$ show a small He effusion peak or shoulder near 400°C, and an effusion structure near 700°C is visible in all cases. While this latter structure is attributed to film crystallization [1], the nature of the 400°C effusion is not fully clarified. Conceivable is a well-defined interconnected void structure, but we note that this 370- 400°C effusion coincides with that of crystalline Si leaving the possibility that these films have some volume parts with a crystalline Si – like structure. The results of the HW films show the presence of isolated voids up to high substrate temperatures ($T_S = 490°C$).

The evaporated Si films (Fig. 3) show again two major He effusion stages below and above about 600°C which we explain in a similar way as for PD and HW material. A minor He effusion near 400°C similar as in HW material is also found. For these evaporated films, different to PD and HW material, a He dose dependence of the LT He effusion peak is observed which may indicate that a dense material with regard to He diffusion forms only after He implantation and may not be present in the as-deposited state. More work appears necessary. The clearly changed He effusion for $T_S = 500°C$ evaporated material is attributed to the fact that $T_S \leq 400°C$ material is amorphous while $T_S = 500°C$ material was found to be crystalline by Raman measurements.

The results for the undoped Si:H films are compiled in Fig. 4. It is seen in the upper part (Fig. 4a) that for the hydrogenated (PD and HW) films, T_{He}^{LT} generally increases with rising T_S up to $T_S \approx 400°C$ thus indicating that a (likely H related) interconnected void structure

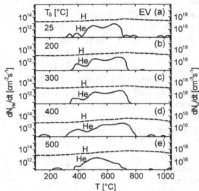

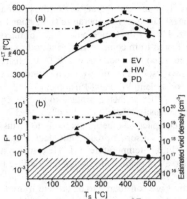

Figure 3. Effusion rate of H and implanted He (dose: 3×10^{15} cm^{-2}) for evaporated (unhydrogenated) a-Si deposited at ((a)-(e)) various substrate temperatures T_S. Note that H effusion rate exceeds barely the background signal.

Figure 4. Temperature T_{He}^{LT} of low temperature helium effusion maximum (a) and parameter F* as well as estimated void density (b) of (undoped) evaporated, HW deposited and plasma-deposited a-Si:H. Hatched range is below detection limit.

disappears at decreasing H concentration and that the density increases. For the evaporated films, T_{He}^{LT} is high even at low T_S indicating that little interconnected voids exist in this material. Apparently, increase of T_S causes only slight changes in this case apart from film crystallization at $T_S > 400°C$. In Fig. 4b, the parameter F* is shown versus T_S. F* is defined by the ratio of HT and LT He effusions and gives the probability that He is trapped in isolated voids on the way to the film surface. In previous work [5] it was estimated that F* =1 equals (roughly) a void concentration of 1.7×10^{19} cm^{-3}. Values of the estimated void density are given on the right hand abscissa. Note that an estimated void density of 10^{17}cm^{-3} (F* $\approx 3 \times 10^{-3}$) is the measuring limit of our effusion apparatus, i.e. values of this order in Figs. 4b and 6a may actually be smaller. The results show high void densities of $10^{19} - 10^{20}$ cm^{-3} for the evaporated films as long as they are amorphous ($T_S < 500°C$) as well as for the HW films. For the PD films, the isolated void concentration in the $T_S \approx 200°C$ films is about 4×10^{18} cm^{-3} at highest, and lower at higher and lower T_S. We note that for device-grade PD films deposited at Bielefeld, even lower F* and void density values (F* = 0.09; estimated void density: 2×10^{18} cm^{-3}) are found [10]. Note that from the molecular hydrogen data [3] a similar concentration of isolated voids can be estimated.

The effusion spectra are changed considerably by doping. In Fig. 5a, hydrogen and (implanted) He effusion is shown for undoped (PD) material (see Fig. 1a), for a-Si:H doped with 1 % diborane (B_2H_6) (Fig. 5b) and for a-Si:H doped with 1 % phosphine (PH_3) (Fig. 5c). The significantly reduced temperature of the H effusion peak for boron doping has been attributed to a Fermi level dependence of the H diffusion energy which is particularly strong on the p – side [1, 11]. The LT He effusion temperature is similar in doped and undoped material [4] but there is a strong doping effect on the height of the HT He effusion peak.

The results of F^* versus T_S for doped PD material are compiled in Fig. 6a. Rather high F^* values near 1-10 corresponding to estimated void concentrations of 10^{19} -10^{20} cm^{-3} are observed near $T_S \approx 200°C$. These values are about 1- 2 orders of magnitude higher than for undoped material. The reduced void concentration at $T_S < 200°C$ is again attributed to the presence of interconnected voids as low temperature H effusion shows up, too. The drop of F^* at high T_S seems to be connected to the decrease in H concentration, as shown in Fig. 6b. These

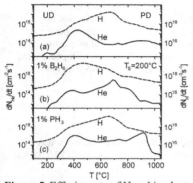

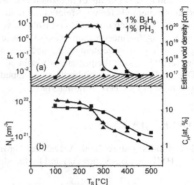

Figure 5. Effusion rate of H and implanted He (dose: 10^{16} cm^{-2}) of (a) undoped, (b) 1% diborane and (c) 1% phosphine doped plasma deposited a-Si:H using a substrate temperature of T_S = 200°C.

Figure 6. Parameter F^* and estimated void density of PD a-Si:H doped with 1% diborane and phosphine (a) and hydrogen density N_H and hydrogen concentration c_H (b) versus T_S. Hatched range is below detection limit.

results suggest that the formation of isolated voids in the (highly doped) PD material requires a hydrogen concentration exceeding about 5 at. %. The different decrease of H density N_H for B- and P-doped material with rising T_S could then cause the isolated voids to disappear at different T_S (see Fig. 6a). Note that the difference in N_H for B- and P-doped a-Si:H (Fig. 6b) can be attributed primarily to the doping (Fermi level) dependence of hydrogen release (surface desorption/out-diffusion) during film growth [1].

Although the nature of the isolated voids in a-Si:H is not fully clarified, the effects of doping on void concentration give some hints. Clearly, the wide variation of F^* and void concentration with doping while the H concentration is only slightly changed (T_S = 200°C PD a-Si:H)) demonstrates that these voids are not directly connected to H concentration. On the other hand, as seen in Fig. 6, a H concentration ≥ 5 at.% seems to be a prerequisite for isolated void formation in doped PD a-Si:H. Conceivable is that voids form in the growth zone by precipitation of molecular hydrogen when the H concentration exceeds about 5 at.% and at substrate temperatures where H diffusivity is high. Transformation of interconnected into isolated voids [4] in the growth zone may also play a role. However, isolated voids and microstructure may also have other (film growth related) origin. We note that the measured high void concentrations for doped PD a-Si:H could account for high measured defect concentrations [12, 13], i.e. a correlation between void and dangling bond concentrations is conceivable for doped a-Si:H. For undoped device grade material, estimated void concentrations near 10^{18} cm^{-3}

exceed the defect density [12, 14] by several orders of magnitude which would mean that most void surfaces are well passivated.

CONCLUSIONS

Effusion measurements of hydrogen and implanted helium are demonstrated to be useful characterization methods for void-related microstructure in thin film silicon materials. Information on both interconnected and isolated voids is obtained. The results show that an increase in hydrogen concentration in a-Si:H causes the diffusion-related He effusion peak temperature to drop in agreement with a decrease in density and the formation of interconnected voids. Our results suggest the presence of isolated voids of the order 10^{18} cm^{-3} in undoped device grade PD a-Si:H and at much higher level in doped PD a-Si:H as well as in undoped evaporated and in non-device grade HW a-Si materials.

ACKNOWLEDGEMENTS

The authors wish to thank A. Dahmen for the ion implantations. Part of the work was financed by BMU (Globe-Si project) and by the state of Nordrhein-Westfalen (project EN/1008B "TRISO").

REFERENCES

1. W. Beyer, F. Einsele, in *Advanced Characterization Techniques for Thin Film Solar Cells*, edited by D. Abou-Ras, T. Kirchartz, U. Rau (Wiley-VCH, Weinheim, Germany, 2011) p. 449.
2. W. Beyer in *Thin-Film Silicon Solar Cells*, edited by A. Shah (CRC Press, Boca Raton, FL, 2010) p. 64.
3. Y.J. Chabal, C.K.N. Patel, *Reviews of Modern Physics* **59**, 835 (1987).
4. W. Beyer, *Phys. Status Solidi* (c) **1**, 1144 (2004).
5. W. Beyer, W. Hilgers, P. Prunici, D. Lennartz, *J. Non-Cryst. Solids* **358**, 2023 (2012).
6. L. Zanzig, *Berichte des Forschungszentrums Jülich* **3087** (Jülich, Germany, 1995).
7. A. Van Wieringen and N. Warmoltz, *Physica* **22**, 849 (1956).
8. O.L. Anderson, D.A. Stuart, *J. Am. Ceram. Soc.* **37**, 573 (1954).
9. G.F. Cerofolini, F. Corni, S. Frabboni, C. Nobili, G. Ottaviani, R. Tonini, *Materials Science and Engineering* **R27**,1 (2000).
10. W. Beyer, W. Hilgers, D. Lennartz, F. Pennartz, P. Prunici, *MRS Symp. Proceedings* **1426**, 341 (2012).
11. W. Beyer, *Physica B* **170**, 105 (1991).
12. D.V. Lang, J.D. Cohen, J.P. Harbison, *Phys. Rev. B* **25**, 5285 (1982).
13. R.A. Street, *Hydrogenated amorphous silicon* (Cambridge University Press, Cambridge, 1991) p. 147.
14. H.Schade in *Thin-Film Silicon Solar Cells*, edited by A. Shah (CRC Press, Boca Raton, FL, 2010) p. 269.

Mater. Res. Soc. Symp. Proc. Vol. 1536 © 2013 Materials Research Society
DOI: 10.1557/opl.2013.751

Temperature Dependence of 1/f Noise and Electrical Conductivity Measurements on p-type a-Si:H Devices

V. C. Lopes, E. Hanson, D. Whitfield, K. Shrestha, C. L. Littler, and A. J. Syllaios
University of North Texas, 1155 Union Circle #311427, Denton, TX 76203-5017

ABSTRACT

Noise and electrical conductivity measurements were made at temperatures ranging from approximately 270°K to 320°K on devices fabricated on as grown Boron doped p-type a-Si:H films. The room temperature 1/f noise was found to be proportional to the bias voltage and inversely proportional to the square root of the device area. As a result, the 1/f noise can be described by Hooge's empirical expression [1]. The 1/f noise was found to be independent of temperature in the range investigated even though the device conductivity changed by a factor of approximately 4 over this range. Conductivity temperature measurements exhibit a $T^{-0.25}$ dependence, indicative of conduction via localized states in the valence band tail [2,3]. In addition, multiple authors have analyzed hole mobility in a-Si:H and find that the hole mobility depends on the scattering of mobile holes by localized states in the valence band tail [4-7]. We conclude that the a-Si:H carrier concentration does not change appreciably with temperature, and thus, the resistance change in this temperature range is due to the temperature dependence of the hole mobility. Our results are applicable to a basic understanding of noise and conductivity requirements for a-Si:H materials used for microbolometer ambient temperature infrared detection.

INTRODUCTION

Hydrogenated amorphous silicon (a-Si:H) is a base material for microelectromechanical systems (MEMS) uncooled infrared imaging focal plane array systems using microbolometer technology [8]. Infrared detection is based on changes in the electrical conductivity with the figure of merit being the temperature coefficient of resistance (TCR). However, electrical 1/f noise may limit detectivity of this material technology. The 1/f noise is attributed to charge number (McWhorter model) or mobility fluctuations (Hooge model) in the conductivity [1]. In a-Si:H material, 1/f noise has been correlated to hydrogen content and microcrystallinity [9], and generation-recombination noise [10]. In addition, authors have related 1/f noise with variable range hopping [11,12]. In this paper, we report on electrical conductivity and 1/f noise measurements made on a-Si:H material.

EXPERIMENTAL

Amorphous silicon thin films were grown by the plasma enhanced chemical vapor deposition (PECVD) technique. Samples were p-type by Boron doping. Measurements were made on two samples: S-30 (H_2 dilution of SiH_4 16:1, BCl_3 to SiH_4 ratio 0.30, and thickness 505 Å), and S-192 (H_2 dilution of SiH_4 55:1, BCl_3 to SiH_4 ratio 0.32, and thickness 1060 Å). XPS measurements show a total concentration of ~ 10% atomic concentration of the highest BCl_3 to SiH_4. Test structures were fabricated using standard photolithographic techniques.

The total measured noise is comprised of the noise generated by the sample and the Johnson noise due to the feedback resistor, (1a). The amplifier gain is determined by the ratio of the amplifier output voltage to the device input or bias voltage, (1b). This gain is then used to determine the device resistance as well as to adjust the measured noise for the device or sample noise.

$$V_m^{\,2} = V_{JR_{Fbk}}^{\,2} + (V_n G)^2 \qquad G = \frac{V_{out}}{V_{in}} = \frac{R_{Fbk}}{R} \qquad \text{(1a) and (1b)}$$

The measured noise is comprised of 1/f noise of the device and white noise due to the device and the feedback resistor.

For the device, the electrical noise is comprised of a white noise component, usually Johnson noise, and of low frequency excess or 1/f noise. The total noise is determined in quadrature

$$V_n^{\,2} = V_{1/f}^{\,2} + V_{Johnson}^{\,2} \tag{2}$$

where $V_{1/f}$ is the 1/f noise voltage and $V_{Johnson}$ is the Johnson noise voltage. The Johnson noise is determined by the sample resistance and temperature. Quantitatively,

$$V_{Johnson}^{\,2} = 4k_B TR(T)\Delta f \qquad R(T) = \rho(T)\frac{\ell}{A} \qquad \text{(3a) and (3b)}$$

The low frequency noise can be described by the Hooge expression [1]

$$V_{1/f}^{\,2} = \frac{\alpha_H V_{Bias}^{\,2}}{Nf} = \frac{B}{f} \tag{4}$$

where f is the frequency, α_H is the Hooge parameter, N is the total number of carriers in the device ($N = P_{Tot} A d$) and V_{Bias} is the bias voltage. For fixed device (area), the parameter B depends on the bias voltage (5a); and for fixed bias voltage, the parameter $B/V_{Bias}^{\,2}$ depends on the inverse device area (5b).

$$B = \left(\frac{\alpha_H}{N}\right)V_{Bias}^{\,2} \qquad \frac{B}{V_{Bias}^{\,2}} = \left(\frac{\alpha_H}{P_{Tot}d}\right)\frac{1}{A} \qquad \text{(5a) and (5b)}$$

where P_{tot} is the total carrier density. Estimates of the Hooge parameter for a-Si:H material are 1×10^{-3} to 4×10^{-3} [13,14].

RESULTS AND DISCUSSION

An example of a-Si:H noise measurements as a function of bias voltage is shown in Figure 1. Measurements were made on sample S-30 with the device having an area of 30 µm x

30 μm at room temperature. Figure 1 (a) shows the noise measurement as a function of frequency for various bias voltages. For frequencies less than100Hz, the noise voltage is observed to increase with decreasing frequency and the noise voltage increases with increasing bias voltage. For frequencies greater than 100 Hz, the noise voltages do not depend on bias voltage. Figure 1 (b) shows the 1/f noise as a function of frequency for various bias voltages.

The 1/f noise in Figure 1 (b) was fit by regression analysis. The noise in this region is observed to depend as ~ $1/f^n$ with n near 0.5 and to increase with increasing bias voltage. Figure 2 (a) shows $\sqrt{B}$ as a function of V_{Bias}. This data was fit to determine the slope which is related to the Hooge parameter. For this device, $P_{Tot}/\alpha_H \approx 2 \times 10^{21} cm^{-3}$. Noise measurements as a function of device area were also made on sample S-30. The device areas measured were 30 μm x 30 μm, 50 μm x 50 μm, and 30 μm x 300 μm. Figure 2 (b) shows B/V_{Bias}^2 as a function of the inverse device area. The parameter B/V_{Bias}^2 is observed to depend linearly with the inverse of the device area. These results (Figures 2 (a) and (b)) show that noise in a-Si:H can be described by the Hooge model.

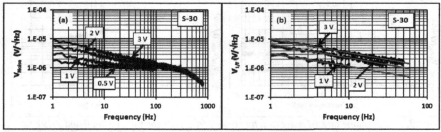

Figure 1. (a) Noise and (b) 1/f noise as function of frequency for various bias voltages.

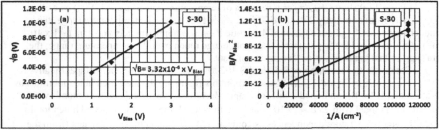

Figure 2. (a) $\sqrt{B}$ as a function of V_{Bias} and (b) B/V_{Bias}^2 as a function of inverse device area.

Noise measurements were made as a function of temperature. The device dimensions are as follows; S-30: 30 μm x 300 μm, and S-192: 50 μm x 1000 μm. Graphs in Figure 3 show the noise voltage and 1/f noise for S-30; and the noise voltage and 1/f noise for S-192 for temperatures ranging from approximately 270 °K and 320 °K. For frequencies greater than 70

Hz for S-30 and 30 Hz for S-192, the noise voltage is lower at the higher temperature. In this frequency range, the noise voltage is due to the sample's Johnson noise and the system noise (white noise). The difference in the measured noise voltage with temperature is attributed to the change in the white noise with temperature. For frequencies less than 10 Hz, the 1/f noise spectra of S-30 and S-192 show little temperature dependence.

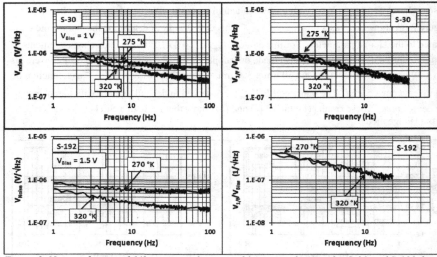

Figure 3. Noise voltage and 1/f noise as a function of frequency for samples S-30 and S-192 for various temperatures.

Figure 4 (a) shows P_{Tot}/α_H as a function of temperature (specifically $1/T^{0.25}$) as well as conductivity from nominally 270 °K to 325 °K for S-192. Over this temperature range, the conductivity varies by about a factor of 4.3 while P_{Tot}/α_H is approximately constant ($\sim 1 \times 10^{21}$ cm^{-3}). Provided the Hooge parameter does not change much over this temperature range, the data suggest that the total carrier density is constant.

Conductivities for samples S-30 and S-192 as a function of temperature (specifically $1/T^{0.25}$) are shown in Figure 4 (b). Also shown in this figure are the measured hole mobilities by Tiedje et al. [4] for Boron doped a-Si:H (1.7 ppm Boron) and the calculated hole mobility using the Schiff [7] model with $L/E = 2 \times 10^{-9} \ cm^2/V$. In general, the conductivity and hole mobility have a similar temperature dependences as compared to the total number of carriers (from the 1/f noise data). The conductivity follows a $1/T^{0.25}$ dependence and is attributed to Mott variable range hopping [2,3]. Here, conduction is due to electrons tunneling to/from localized states in the valence band tail. Multiple authors [4-7] have determined that the hole mobility is determined by trap states in the valence band tail. Both the conductivity as well as the hole mobility are related to the localized states within the valence band tail.

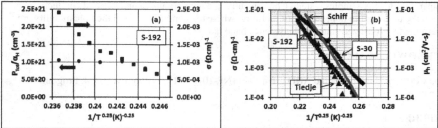

Figure 4.(a) Noise and conductivity data as a function of temperature for sample S-192. (b) Conductivity of S-30, S-192, the hole mobility in a-Si:H as reported by Tiedje et al. [4] and the calculated hole mobility using the Schiff model [7]as a function of temperature.

In order to understand the temperature dependent noise data, a sensitivity anaylsis was done. In our material, the total number of charge carriers includes hopping electrons and mobile holes. In this analysis, we assume that the total number of charge carriers is fixed. Given the high Boron content in the film, we assume that the number of mobile holes is constant over the 270 to 325 °K temperature range, which in turn implies that the number of hopping electrons is also constant over this temperature range. For our sensitivity analysis, the Hooge parameter is taken to be constant with $\alpha_H = 2 \times 10^{-3}$, and the total charge carrier density was determined from $P_{Tot}/\alpha_H \approx 2 \times 10^{21} cm^{-3}$. The calculated noise was determined by (1) though (5) using $V_{Bias} = 1$ V. Figure 5 shows the calculated total noise and 1/f noise, respectively, at temperatures of 275 K and 320 K. It is seen that, at low frequencies (less than 10 Hz), 1/f noise dominates the noise spectrum, and at high frequencies (greater than 60 Hz), the white noise dominates the noise spectrum, similar to that seen in the experimental results shown in Figure 3.

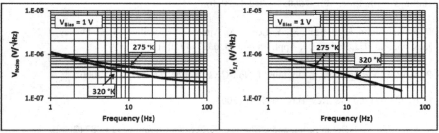

Figure 5. The calculated noise voltage and 1/f noise as a function of frequency.

CONCLUSIONS

Noise measurements were made on Boron doped p-type a-Si:H films. The 1/f noise can be described by the Hooge model. The 1/f noise was observed to be relatively independent of temperature over the temperature range tested. Provided the Hooge parameter does not change with temperature, our data suggest that the total number of charge carriers does not vary with temperature for the samples tested. Consequently, the temperature dependence of conductivity is due to the temperature dependence of mobility, where the conductivity and the hole mobility are determined by localized states in the valence band tail.

ACKNOWLEDGMENTS

This work was done under the ARO grant W911NF-10-1-0410, William W. Clark Program Manager.

REFERENCES

1. 1. F. N. Hooge, IEEE Trans Elect Dev **41**, 1926 (1994).

2. A. Lewis, Phys. Rev. Lett. **29**, 1555 (1972).

3. N. Savvides, J. Appl. Phys. **56**, 2788 (1984).

4. T. Tiedje, J. M. Cebulka, D. L. Morel, and B. Abeles, Phys. Rev. Lett. **46**, 1425 (1981).

5. S. Dinca, G. Ganguly, Z. Lu, E. A. Schiff, V. Vlahos, C. R. Wronski, and Q. Yuan, Mat. Res. Soc. Symp. Proc. **762**, A7.1.1 (2003).

6. M. Brinza and G. J. Adriaenssens, J. Optoelectronics and Advanced Materials **7**, 73 (2005).

7. E. A. Schiff, Mat. Res. Soc. Symp. Proc. **1153**, 1153-A15-01 (2009).

8. S. K. Ajmera, A.J. Syllaios, Gregory S. Tyber, Michael F. Taylor, and Russell E. Hollingsworth, Proc. SPIE **7660**, 766012-1 (2010).

9. D. B. Saint John, H.-B. Shin, M.-Y. Lee, S. K. Ajmera, A. J. Syllaios, E. C. Dickey, T. N. Jackson, and N. J. Podraza, J. Appl. Phys. **110**, 033714 (2011).

10. P. A. W. E. Verleg and J. I. Dijkhuis, J. Non-Cryst. Solids **266-269,** 232 (2000).

11. B. I. Shklovskii, Phys. Rev. B **67**, 045201 (2003).

12. V. I. Kozub, S. D. Baranovskii, and I. Shimak, Solid State Commun. **113** 587 (2000).

13. K. Kim and R. E. Johanson., IEEE CCECE/CCGEI, Saskatoon, 2261 (2005).

14. J. Rhayem, D. Rigaud, M. Valenza, N. Szydlo, and H. Lebrun., Solid-State Electronics **43**, 713 (1999).

Mater. Res. Soc. Symp. Proc. Vol. 1536 © 2013 Materials Research Society
DOI: 10.1557/opl.2013.614

Study of Surface Passivation of CZ c-Si by PECVD a-Si:H Films; A Comparison Between Quasi-Steady-State and Transient Photoconductance Decay Measurement

Omid Madani Ghahfarokhi, Karsten von Maydell and Carsten Agert
NEXT ENERGY · EWE Research Centre for Energy Technology at Carl von Ossietzky
University Oldenburg, Carl-von-Ossietzky-Str. 15, 26129 Oldenburg, Germany

ABSTRACT

We have investigated the passivation of low lifetime non-polished Czochralski (CZ) mono-crystalline silicon (c-Si) wafers by hydrogenated amorphous silicon (a-Si:H), deposited by plasma enhanced chemical vapor deposition (PECVD) technique. The dependence of the effective lifetime (τ_{eff}) on the deposition parameters including hydrogen gas flow, power and temperature has been studied. Minority carrier lifetime was measured as deposited and also after an annealing step in both quasi-steady-state (QSS) and transient mode of photoconductance decay. By comparison between τ_{eff} measured in each of the aforementioned modes, two distinguishable behaviors could be observed. Moreover, to get further insight into the surface passivation mechanism, we have modeled the recombination at a-Si:H/c-Si interface based on the amphoteric nature of dangling bonds. The results of our modeling show that the discrepancy observed between QSS and transient mode is due to the high recombination rate that exists in the bulk of defective CZ wafer and also partly related to the different thicknesses monitored in each mode. So, by comparison between the injection level dependency of τ_{eff} measured in QSS and transient modes, we introduce a valuable technique for the evaluation of c-Si bulk lifetime.

INTRODUCTION

To determine the minority carrier lifetime, several techniques such as infrared (IR) transmission or emission, photoluminescence (PL), eddy-current and photoconductance (PC) have been developed. Among them, PC technique, developed by Sinton Instruments, is the most widely used technique since it is contactless, fast and the obtained results have low uncertainty [1]. In the PC technique, the measurements can be performed in two different modes: QSS and transient mode. As shown by Nagel [2] the lifetime can be calculated via equation 1:

$$\tau_{eff} = \Delta n/(G-(d\Delta n/dt)) \tag{1}$$

Where G is generation rate [cm^{-3}s^{-1}] and Δn is the excess charge carrier density [cm^{-3}] . In the QSS mode, the decay of flash light is at least 10 times slower than the carrier lifetime. Thus, the excess carrier populations are always in the steady-state condition, or very close to it. So the lifetime (τ) can be calculated via equation 2:

$$\tau_{eff} = \Delta n/G \tag{2}$$

In transient mode, the initial generation is done by a very short and intensive light pulse and afterwards there is no generation. Therefore, the lifetime (equation 1) is reduced to equation 3:

$$\tau_{eff} = -\Delta n/(d\Delta n/dt) \qquad (3)$$

In some cases, a discrepancy is observed between the τ_{eff} acquired in each mode [3,4]. To our knowledge, the source of this discrepancy is not well investigated. Moreover, we believe this discrepancy could be used as a valuable source to obtain additional information about the c-Si bulk life time which will be addressed in this work.

The first part of this report focuses on the passivation of low lifetime non-polished CZ wafers by a-Si:H layers. We address the effect of plasma parameters namely, hydrogen flow, power and temperature on the passivation of a-Si:H films. Afterward, for more in-depth understanding of the passivation mechanism, the recombination at a-Si:H/c-Si interface based on the amphoteric nature of dangling bonds is modeled. Then, based on the modeling results, the discrepancy between QSS and transient mode is well explained. Finally, finding the source of discrepancy enables us to gain additional information about the quality of the bulk c-Si wafers.

EXPERIMENTAL DETAILS

For the passivation study, n-type non-polished CZ wafers with resistivity and thickness in the range of 4-8 Ω·cm and 180±30 µm were used, respectively. Furthermore, we have only performed a saw damage removal by wet chemical solution in order to remove the kerf damage. It should be noted that no additional polishing or texturing was performed. To remove native oxide from c-Si surface, prior to the deposition of each passivation layer, we immersed the substrates in HF solution (1%) for 2 minutes. For deposition of a-Si:H layers, parallel plate PECVD reactor operated at 13.56 MHz was used. The intrinsic a-Si:H layers with the thickness of 18±3 nm were deposited on both sides of the wafers. Effective minority carrier lifetimes of the samples were evaluated with Sinton Consulting WCT-120. The post-deposition annealing was done at 190 °C for 60 min in air ambient.

RESULTS AND DISCUSSION

1-Effect of hydrogen flow, power and temperature

Figure 1a shows the measured τ_{eff} of c-Si wafers, passivated by a-Si:H with various H_2 flows in the range of 0-300 sccm. Two series of samples deposited with power of 15 and 25 W are demonstrated. The values for τ_{eff} were taken at minority carrier density (Δn) of 10^{15} cm^{-3}. As shown in figure 1a, the measured τ_{eff} values before the annealing process do not show any significant improvement when the H_2 flow is increased. The measured τ_{eff} values only change from 16 to 101 µs and 16 to 58 µs for deposition power of 15 and 25 W, respectively. After a post-deposition annealing, two significantly different τ_{eff} behaviors could be observed. For 15 W series, samples deposited with 100 sccm (R=[H$_2$]/[SiH$_4$]=2.5) or less H_2 flow show significant improvement in the τ_{eff}. Similarly, the same improvement trend was observed for 25 W series deposited with H_2 flow up to 200 sccm (R=5). However, above 100 sccm for 15 W series (200 sccm for 25 W series), QSS lifetime value was less than as-deposited value. In a study by De Wolf [5] it has shown that the drop in the QSS lifetime after post-deposition annealing is a good indication for the epitaxial growth. Hence, we could attribute this reduction of QSS lifetime value (after annealing step) to the formation of epitaxial growth, due to the high hydrogen dilution. Also, the difference between 15 and 25 W sample series could be related to the higher silane depletion fraction with increasing power [6]. This could shift the onset of epitaxial growth for the case of 25 W to higher hydrogen dilution (as drop in QSS lifetime after post- deposition annealing was observed at higher hydrogen dilution) compared to 15 W sample series. It should be noted that the drop in QSS lifetime after post-deposition annealing (an indication of epitaxial growth) happens at the same hydrogen dilution for both polished and non-polished wafers. This could indicate that the micrometer roughness of the non-polished wafers does not have any effect on the onset of the epitaxial growth.

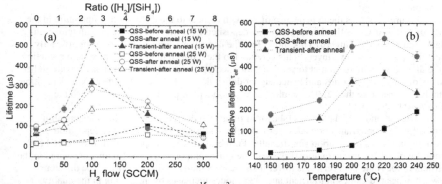

Figure 1 (a) The variation of the τ_{eff} (Δn: 10^{15} cm^{-3}) with respect to H_2 flow for two different powers of 15 and 25 W. The measurement was done as-deposited and also after a post-deposition annealing in both QSS and transient mode. (b) The τ_{eff} (Δn: 10^{15} cm^{-3}) versus the deposition temperature. The respective square and circle data series correspond to the QSS lifetime before and after the annealing process. The triangle data series shows the transient mode lifetime after the annealing process.

For samples with the epitaxial growth, we have measured a higher lifetime in the transient than QSS mode. Under transient conditions, a sample with high recombination surfaces (like epitaxial growth) will have a larger proportion of carriers near the center of the wafer after a very short period of time, since those carriers recombining at the surfaces are not rapidly replenished by newly-generated carriers. Consequently and roughly speaking, while there is not any charge carrier at the interface, a reduction of carriers ($d\Delta n/dt$) exists in the bulk. This means that we have measured the recombination rate close to the center of wafer rather than the interface (see equation 3). For the QSS mode, the existing generation could compensate some parts of the recombination at epitaxial interface. However, due to the much faster recombination rates at the epitaxial interface, the lifetime drastically decreases. Unlike the epitaxial samples, the post-deposition annealing improves the passivation due to the restructuring of the amorphous network. By formation of a new network with more bonding states and reduced strained Si-Si bonds, the defect density in a-Si:H network is reduced [7].

From figure 1a, it could be seen that by increasing the H_2 flow (prior to the epitaxial growth) the lifetime increases. The increase of the lifetime is related to the dependence of the a-Si:H film microstructure and the a-Si:H/c-Si interface on the plasma hydrogen content [8].

Deposition of a-Si:H at higher hydrogen flow (prior to epitaxial growth) leads to lower dangling bonds in a-si:H network and hence higher passivation quality [9]. As a general trend, lower τ_{eff} was measured for 25 W series compared to 15 W one, which is attributed to the damage of the interface with higher ion bombardment at higher power [10].

For both power series samples without any epitaxial growth, higher lifetimes in QSS mode than in transient mode were measured. By optimizing the quality of the passivation layer, the increase in the QSS lifetime is higher than transient lifetime. We attribute this higher QSS lifetime first to the high recombination rate that exists in the bulk of low quality used CZ c-Si wafers and second to the different monitored thicknesses in each mode. This is further explained in section 2.

Figure 1b shows the dependence of τ_{eff} with respect to the deposition temperature in the range of 150-240°C. As the data suggests, as-deposited QSS lifetime increases with temperature (black data series in figure 1b).This shows that the deposition temperature could significantly affect the quality of grown layers. This is related to the reduced mobility of spices and reduction in the diffusivity of hydrogen at lower deposition temperatures which results a layer with higher density of unsaturated bonds and hence higher defect density [11]. On the other hand, deposition at higher temperature than 220°C, could also result in a poor passivation layer, due to the effusion of hydrogen [12].

The red and blue data series (shown in figure 1b) demonstrate τ_{eff} after the annealing process for QSS and transient modes, respectively. As already discussed, post-deposition annealing improves passivation significantly. Furthermore, more improvement is observed for samples with higher initial quality. Likewise the other sample series, the lifetime measured in the transient mode (blue data series) is lower than the QSS mode (red data series). The difference between measured values in transient and QSS modes increases by increasing in the quality of passivation layer.

2-Modeling results

To study the mechanism of passivation by a-Si:H layers and obtain an insight into the discrepancy between the QSS and the transient mode, we have theoretically modeled the recombination at the a-Si:H/c-Si interface. As the recombination at a-Si:H/c-Si interface takes place through dangling bonds, we have used the closed-form recombination model, introduced by J. Hubin [13]. This model allows us to distinguish between two approaches of passivation, either field effect passivation by surface charge density (Q_s) or reduction of interface defect density (N_s). The details description of this modeling is beyond the scope of this study and will be reported elsewhere. It should be mentioned that the obtained values for N_s and Q_s by our modeling are not absolute values, however they could be used for relative comparison between various samples. In our modeling the neutral electron to hole capture cross section and the ratio of charged to neutral capture cross section were set to 0.05 and 550, respectively. The neutral hole capture cross section was assumed to be 10^{-16} cm^2 [4].

Figure 2a shows the injection level dependent of τ_{eff} for samples deposited at various temperatures. The solid and dashed lines represent the modeling data whereas the data points illustrate the experimental results. As figure 2a suggests, the experimental data points are well fitted with the modeling results. Figure 2b demonstrates the N_s, extracted from the modeling with respect to the deposition temperature for both QSS and transient mode. From figure 2b, it could be seen that there is a direct relation between the improvement in the passivation and reduction in N_s. Hence, the passivation by intrinsic a-Si:H should be a chemical passivation. Besides, comparison between the injection level dependent of the τ_{eff} acquired in QSS (filled data points) and transient mode (open data points), shows that the τ_{eff} is lower in transient mode for all the injection levels, as shown in figure 2a. This could be related to the lower thickness of passivation layer, monitored in transient mode (compared to the QSS mode). Although, there is no detailed knowledge on the depths of passivation layer that recombination takes place, here interestingly, the modeling results suggest a higher N_s in transient than QSS mode (see figure 3b). These results confirm the hypothesis that the interface layer monitored in transient mode is thinner than QSS mode. As in the transient mode, the most defective part of the passivation layer (first few nanometers layer) is analyzed, hence a lower τ_{eff} is measured. On the other hand in the QSS mode, a thicker layer of the passivation is monitored which results in higher τ_{eff}.

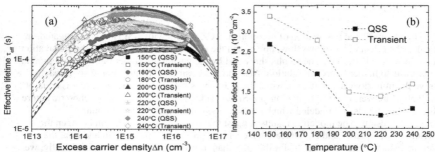

Figure 2 (a) The injection level dependent of τ_{eff} for samples deposited at various temperatures after post-deposition annealing. The data points represent the experimental results and solid and dashed lines illustrate simulation data. (b) Interface defect density (N_s) variation with respect to the deposition temperature, extracted from the modeling results for both QSS and transient mode.

The bulk recombination could also cause this discrepancy. Considering the principle that exists behind the QSS and transient modes, bulk recombination has a stronger influence on the transient lifetime. As shown in equation 3 for the transient mode, there is an inverse relation between the lifetime and reduction rate of excess charge carrier density ($-d\Delta n/dt$). As a result, if the recombination rate is high (such as low lifetime bulk wafers or low passivation quality), its influence on the transient lifetime is much higher than the QSS mode. Also it should be considered that in the transient mode, the recombined carriers cannot be replenished (at least for low and medium injection level). Therefore, the effect of the different monitored thicknesses on the τ_{eff} is more significant at low and medium injection levels and the effect of high recombination in the bulk could be observed for all the injection levels (and exclusively for high injection levels). In figure 2a, as the acquired τ_{eff} in the transient mode shows significantly lower values in whole injection levels than QSS mode, thus we believe that the effect of recombination in the bulk is more than the effect of different thicknesses, monitored in each mode. Another reason that confirms this hypothesis is higher improvement for the QSS lifetime than the transient lifetime, by optimizing the quality of the passivation layers (see figure 1a and b). By improving the passivation layer, it is expected that the difference between QSS and transient lifetime reduces. However, due to the higher impact of the bulk recombination on the transient lifetime (compared to the QSS lifetime), higher improvement for the QSS lifetime (especially at medium injection levels) was obtained (see figure 2a).

Finally, this result indicates that comparison between the acquired τ_{eff} in the QSS and transient modes could be a valuable tool for examining the quality of the c-Si wafers. So for high bulk lifetime c-Si wafers and high quality passivation layers, only a few differences in the QSS and transient lifetimes at low and medium injection levels are expected to be observed. For high bulk lifetime c-Si wafers and poor passivation layers, the difference in the QSS and transient lifetime increases at low and medium injection levels and at high injection levels similar (almost the same) lifetime values are expected. For the case of low bulk lifetime c-Si wafers, the difference in the QSS and transient modes is much more and could be observed for the whole injection levels.

CONCLUSIONS

In summary, the passivation of non-polished CZ wafers with a-Si:H are investigated. In particular, the effect of hydrogen flow, power and the deposition temperature on the passivation of the c-Si wafers is studied. The results show that by increasing the power the onset of epitaxial growth shifts to higher hydrogen dilution. Also, post-deposition annealing could significantly improve the minority carrier lifetime while the initial passivation of deposited layers is equally important. Moreover, the recombination at a-Si:H/c-Si interface based on the amphoteric nature of the dangling bonds are modeled which shows a good agreement with our obtained experimental results. The modeling results suggest that there is a direct relation between the passivation layer quality and N_s, indicating a chemical passivation process. In addition, the modeling results suggest a lower N_s for the QSS than transient measurement mode. Finally, we demonstrate that the comparison between injection level dependent of τ_{eff} acquired in transient and QSS modes is a valuable tool for c-Si bulk lifetime evaluation.

REFERENCES

1. R. Sinton and A. Cuevas, Appl. Phys. Lett. 69, 2510 (1996).
2. H. Nagel, C. Berge and A. G. Aberle, J. Appl. Phys. 86, 6218 (1999).
3. J. Schmidt, IEEE Trans. Electron. Dev. 46, 2018 (1999)
4. S. Olibet, E. V. Sauvain and C. Ballif, Phys. Rev. B 76, 035326 (2007).
5. S. De Wolf and M. Kondo, J. Appl. Phys. 90, 042111 (2007).
6. A. Descoeudres, L. Barraud, R. Bartlome, G. Choong, Stefaan De Wolf, F. Zicarelli, C. Ballif, Appl. Phys. Lett. 97, 183505 (2010).
7. M. Z. Burrows, U. K. Das, R. L. Opila, S. De Wolf and R. W.Birkmire, J. Vac. Sci. Technol. A 26(4), 683 (2008).
8. R. A. Street, Phys. Rev. B 43, 2454 (1991).
9. T. F. Schulze, H. N. Beushausen, C. Leendertz, A. Dobrich, B. Rech and L. Korte, Appl. Phys. Lett. 96, 252102 (2010).
10. M. S. Jeon, K. Kawachi, P. Supajariyawichai, M. Dhamrin and K.Kamisako, e-J. Surf. Sci. Nanotech. 6, 124 (2008).
11. A. Matsuda, M. Takai, T. Nishimoto, M. Kondo, Sol. Energ. Mat. Sol. Cells 78, 3 (2003).
12. D. K. Biegelsen, R. A. Street, C. C. Tsai, and J. C. Knight, Phys. Rev. B 20, 4839 (1979).
13. J. Hubin, A.V. Shah and E. Sauvain. Philosophical Magazine Letters 66, 115 (1992).

Nanostructured Silicon and Related Novel Materials

Mater. Res. Soc. Symp. Proc. Vol. 1536 © 2013 Materials Research Society
DOI: 10.1557/opl.2013.598

Charge transport in nanocrystalline germanium/hydrogenated amorphous silicon mixed-phase thin films

Kent E. Bodurtha and J. Kakalios
School of Physics and Astronomy, University of Minnesota, Minneapolis, MN 55455

ABSTRACT

Mixed phase thin films consisting of hydrogenated amorphous silicon (a-Si:H) in which germanium nanocrystals (nc-Ge) are embedded have been synthesized using a dual-chamber co-deposition system. Raman spectroscopy and x-ray diffraction measurements confirm the presence of 4 - 4.5 nm diameter nc-Ge homogenously embedded within the a-Si:H matrix. The conductivity and thermopower are studied as the germanium crystal fraction X_{Ge} is systematically increased. For $X_{Ge} < 10\%$, the thermopower is n-type (as in undoped a-Si:H) while for $X_{Ge} > 25\%$ p-type transport is observed. For films with $10 < X_{Ge} < 25\%$ the thermopower shifts from p-type to n-type as the temperature is increased. This transition is faster than expected from a standard two-channel model for charge transport.

INTRODUCTION

There has recently been great interest in the electronic properties of mixed-phase thin films consisting of semiconducting nanocrystals embedded within an amorphous semiconductor matrix [1,2]. These materials combine the large-area advantages of amorphous semiconductors with the superior optoelectronic properties of crystals. The charge transport properties of these composite materials can depend sensitively on the nanocrystal concentration [3]. We report here a unique transition in the conduction mechanism in intrinsic mixed-phase thin films of hydrogenated amorphous silicon (a-Si:H) in which nanocrystalline germanium (nc-Ge) particles have been homogenously embedded. The conductivity and thermoelectric effect in the nc-Ge/a-Si:H films are studied as the germanium crystal fraction X_{Ge} is systematically varied from 0% to 75%. A transition from conduction through the a-Si:H matrix to through the nc-Ge phase is reflected in measurements of the thermopower, which finds a change from n-type (for $X_{Ge} < 10\%$) to p-type transport (for $X_{Ge} > 25\%$) as the nc-Ge concentration is increased. For germanium crystal fractions corresponding to the n- to p-type transition ($10 < X_{Ge} < 25\%$), the thermopower is n-type above 400K and p-type near room temperature. The transition from n-type to p-type thermopower is inconsistent with a simple two-channel mode of charge transport.

MATERIALS PREPARATION

The nc-Ge/a-Si:H films described here were synthesized in a dual-chamber co-deposition plasma enhanced chemical vapor deposition (PECVD) system that enables the growth of a wide variety of mixed-phase thin film materials, described in detail previously [4]. Nanocrystalline particles are synthesized in an upstream flow-through tube plasma reactor [5], and are then injected into a second capacitively-coupled plasma (CCP) deposition system in which the surrounding matrix material is grown. The synthesis of the nanocrystals is completely decoupled from the deposition of the host semiconductor matrix, so that the growth conditions can be independently optimized for each component of the mixed-phase material. This deposition system has been completely rebuilt prior to the synthesis of the nc-Ge/a-Si:H films and has not been exposed to either phosphine or diborane. Consequently unintentional doping by contaminant gases is excluded as an explanation for the results described here.

Figure 1 shows a schematic drawing of the dual-chamber co-deposition system. A mixture of germane (GeH$_4$) and argon flow through the upstream nanoparticle synthesis chamber, which consists of 3/8" quartz tube with ring electrodes, for which the deposition conditions (80W, 2.3Torr) are optimized to promote the growth of germanium nanocrystals. The nc-Ge are then entrained by the argon flow and injected into the second CCP reactor, in which a-Si:H is grown from silane in a plasma at a lower power and pressure (4W, 0.25 Torr). The typical film thickness is ~ 1 µm and the films are deposited onto Corning 1737F quartz at a substrate temperature of 250°C. Aluminum electrodes (thickness ~ 100 nm) are evaporated in a co-planar configuration onto the surface of the films. Due to gas convection currents in the CCP chamber, the concentration of nc-Ge in the film decreases as a function of distance from the injection tube that brings the particles in from the particle synthesis chamber. In each deposition run, a series of mixed phase films are grown in which the nanocrystal concentration is varied. Both the nanocrystal diameter and the host a-Si:H matrix are identical, and the only parameter that varies is the concentration of the nanocrystals. The data presented here represent the results from several such deposition runs, each one yielding approximately four to eight samples of varying crystal content.

The presence of nc-Ge within the a-Si:H matrix is confirmed using x-ray diffraction and Raman spectroscopy. Based on analysis of the x-ray diffraction peak widths, the crystallites have diameters of ~ 4 - 4.5 nm. Raman spectra were recorded with a 514.5 nm argon ion laser at a power of ~ 5 mW with care taken to avoid heating of the films. Representative Raman spectra are shown in Figure 2, with some films being omitted for clarity. The sharp feature at 300 cm^{-1} is consistent with the c-Ge TO mode, while the broad peak at 480 cm^{-1} is associated with a TO excitation of the amorphous silicon matrix [6,7]. The broad feature at ~ 400 cm^{-1} has been ascribed to Si-Ge bonds in studies of homogenous a-Ge$_x$Si$_{1-x}$:H alloys [8]. Rutherford backscattering was employed to ascertain the total germanium content of representative mixed-phase films. Calibration of the total germanium content enables the determination of the relative Raman scattering cross-sectional area for nc-Ge, compared to a-Si:H. Details of the structural characterizations will be published separately.

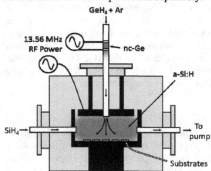

Figure 1: Sketch of the dual-chamber co-deposition plasma enhanced chemical vapor deposition system. Germanium nanocrystals are synthesized in an upstream reactor and injected into a second plasma, where the nc-Ge is embedded in the growing a-Si:H.

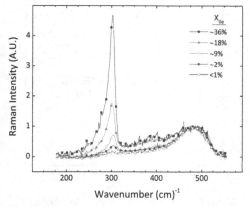

Figure 2: Raman spectra of a series of a-Si:H films containing nc-Ge inclusions. The sharp peak at 300 cm^{-1} reflects the TO mode of c-Ge while the broad peak at 480 cm^{-1} results from the TO mode of a-Si.

RESULTS

A transition from conduction through the a-Si:H matrix to charge transport via interconnected germanium nanocrystals is reflected in measurements of the dark conductivity. Figure 3a shows Arrhenius plots of the conductivity σ of the nc-Ge/a-Si:H films for nc-Ge crystal fraction X_{Ge} ranging from 0% (pure a-Si:H) up to X_{Ge} = 60%. For clarity only some of the conductivity curves have been plotted. These data are well described by the expression $\sigma = \sigma_o \exp[-E_\sigma/k_B T]$, where σ_o is the pre-exponential factor and the conductivity activation energy $E_\sigma \sim 0.75$ eV for the low germanium content films and $E_\sigma \sim 0.4$ eV for higher germanium nanocrystalline concentrations.

Films with $10 < X_{Ge} < 25\%$ display significant curvature in an Arrhenius plot. Shown in Figure 3b is one such measurement that has been fit (dashed line) using the sum of two exponentials with activation energies $E_\sigma \sim 0.75$ (circles) and ~ 0.4 eV (squares), representing a-Si:H and nc-Ge conduction, respectively. In order to gain further insight into the quality of the achieved fit, the "reduced activation energy" is calculated, using a procedure developed by Zabrodskii and Shlimak [9]. The logarithmic derivative of the conductivity $w(T) = d\ln\sigma/d\ln T$ is computed, and plotted against temperature on a log-log graph. If the slope of the resulting plot is -1, this indicates thermally activated conduction, with different activation energies represented as parallel lines. This analysis, calculated from the curves in Figure 3b and shown in the inset, serves as further evidence of the form of the conductivity for these samples.

The change in charge transport in the nc-Ge/a-Si:H films as the germanium crystal fraction is increased is also apparent in measurements of the temperature dependence of the thermoelectric coefficient (the Seebeck effect). In this measurement the sample rests across two separate copper blocks which are 4 mm apart; inside of each Cu block a 50W cartridge heater is embedded. The temperature of each block is controlled independently by a dual-channel temperature controller. The entire measurement system, capable of measuring the electrical properties of high impedance thin films down to $\sigma \sim 10^{-8}$ Ω^{-1}cm^{-1}, resides in a vacuum chamber. The temperatures of the blocks T_1 and T_2 are such that their average is $T_{avg} = (T_1 + T_2)/2$, and as a result of the temperature gradient $\Delta T = T_1 - T_2$, a thermoelectric voltage is induced across the electrodes. After recording this voltage the temperature gradient is adjusted, always maintaining

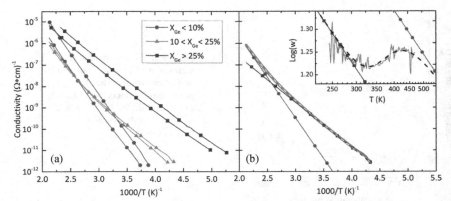

Figure 3: Plot of conductivity against 1000/T (a) for nc-Ge/a-Si:H thin films for which X_{Ge} is increased from 0 (pure a-Si:H) to 60%. Symbols are placed every 18 data points for clarity. (b) Curvature in films with $10 < X_{Ge} < 25\%$ are fit using the sum of two exponentials as indicated by the calculation of the reduced activation energy (inset).

the same average set temperature T_{avg}. For each average temperature, we generate thermal gradients of $\Delta T = \pm 12$ K and 0 and record the respective induced voltage. This procedure eliminates the contribution of any small temperature-dependent voltage offset to the signal. Temperature stability is maintained within ± 0.05 K of the set temperatures. The measured Seebeck coefficient is derived from the slope of the resulting linear plot of the induced voltage against ΔT, and the process is repeated at a new average temperature, from 350K to 450K.

Figure 4a shows plots of the resulting Seebeck coefficient against $1000/T_{avg}$ for films with X_{Ge} ranging from 0% up to 75%. When $X_{Ge} < 10\%$, a negative thermopower is observed; fits of these data to $S = (k_B/e)[E_S/k_B T + A]$ find an activation energy $E_S \sim -0.5$ eV, consistent with previous measurements of undoped a-Si:H [10]. For germanium crystal fractions between approximately $10 < X_{Ge} < 25\%$ the thermopower exhibits a temperature dependent transition to positive values. For $X_{Ge} > 25\%$, the thermopower is positive for all temperatures examined, with a smaller slope of $E_S \sim +0.2$ eV, as observed in studies of thin film crystalline germanium [11]. A discrepancy between the activation energy obtained from the temperature dependence of the thermopower and that found from Arrhenius plots of the dark conductivity is commonly observed in thin film amorphous semiconductors. The difference in the activation energies can be up to several hundred meV and is typically ascribed to the influence of long-ranged disorder, such as potential fluctuations or composition modulations, on electronic transport [12,13]. However, such a difference is not normally observed in crystalline materials, and the difference in activation energies for the $X_{Ge} > 25\%$ films is surprising.

DISCUSSION

One interpretation of the nc-Ge/a-Si:H transport data presented here is that in mixed-phase films with a low concentration of nc-Ge, conduction occurs predominantly through the a-Si:H matrix. The observed conductivity and n-type thermopower is comparable to that typically found in undoped a-Si:H, a result of the much larger mobility for electrons compared to holes [10,14]. For germanium crystal fractions $X_{Ge} > 25\%$, extended, percolating chains of germanium nanocrystals form. These Ge channels have a lower activation energy than the surrounding a-Si:H, and short out the amorphous silicon matrix more strongly at lower temperatures. Previous studies of bulk crystalline germanium have also reported p-type thermopower values [11], but hydrogenated microcrystalline germanium is typically found to be n-type [15]. Our results may differ from Stewart *et. al.* [15] due to differences in crystallite size. Further work is underway to elucidate the mechanism behind this sign reversal.

In this picture the transition arises from the competition between two conduction channels operating in parallel in the film. In this case one should be able to fit the observed temperature dependence of the thermopower using the "two band" expression

$$S(T) = [(1 - X_{Ge})\sigma_{Si} S_{Si} + X_{Ge} \sigma_{Ge} S_{Ge}] / \sigma(T) \qquad (1)$$

where σ_{Si} and σ_{Ge} is the conductivity of the pure a-Si:H phase and the pure nc-Ge phase respectively, S_{Si} and S_{Ge} are the corresponding thermopower values, $S(T)$ is the thermopower as a function of temperature for a given nc-Ge/a-Si:H film in the transition region, and $\sigma(T) = (1 - X_{Ge})\sigma_{Si} + X_{Ge} \sigma_{Ge}$ [14].

Figure 4b shows a plot of the thermopower using the two-band expression above. The values for σ_{Si} and S_{Si}, and σ_{Ge} and S_{Ge} are taken from measurements from the $X_{Ge} = 0$ (pure a-Si:H) and $X_{Ge} = 75\%$ films (high concentration nc-Ge), respectively. While, for $10 < X_{Ge} < 25\%$, a variation of the thermopower results from eq. (1) from that comparable to pure nc-Ge at low temperatures to that similar to pure a-Si:H at higher temperatures, the transition is much more gradual with temperature than actually observed (Figure 4a).

In the transition region, the conductivity data is well-described by the sum of n-type and p-type conduction paths. However, the actual fraction of conduction due to the a-Si:H and nc-Ge for the transition films is not what one would expect from a simple "volume fraction" model as in eq. (1). The fraction of current flowing through the n-type channel can be defined

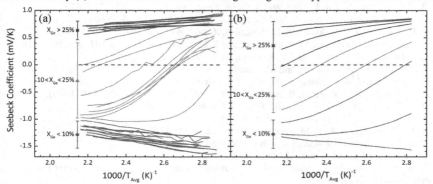

Figure 4: Measured Seebeck coefficients (a) and calculated values from the simple two-band model (b) plotted against $1000/T_{avg}$ for nc-Ge/a-Si:H thin films with $X_{Ge} = 0$ (pure a-Si:H) to $X_{Ge} = 75\%$.

as $\gamma(T) = \sigma_1(T)/\sigma(T)$ where $\sigma_1(T)$ is found from fits to the n-type conductivity as in Figure 3b. It is then possible to calculate thermopower curves that more closely match the data using the expression

$$S(T) = \gamma S_{Si} + (1 - \gamma)S_{Ge} \qquad (2)$$

which assumes that the fractional contribution by each channel is equal in the conductivity and thermopower for a given sample at each temperature. At present, however, it is unclear how γ and X_{Ge} are related. Studies on the dependence of γ on X_{Ge} are underway.

CONCLUSION

We have synthesized novel mixed phase thin films consisting of a-Si:H in which germanium nanocrystals are homogenously embedded, using a two-chamber co-deposition system. X-ray diffraction indicates that the nc are approximately 4 nm in diameter and RBS and Raman spectroscopy provide quantitative determinations of the germanium crystal fraction in these films. The conductivity and thermopower have been studied as the nc-Ge concentration is systematically varied from 0% to 75%. For films with $X_{Ge} < 10\%$, conduction is n-type, similar to that observed in pure a-Si:H, while for $X_{Ge} > 25\%$ transport is p-type, as found in single-crystal Ge. In the transition region $10 < X_{Ge} < 25\%$, transport is n-type above 400K and p-type near room temperature. Further studies are underway to elucidate the transport mechanisms in these novel films ordered on the mesoscale.

This work was partially supported by NSF grant DMR-0705675, the NINN Characterization Facility, the Nanofabrication Center, an Xcel Energy grant under RDF contract #RD3-25, NREL Sub-Contract XEA-9-99012-01, and the University of Minnesota.

REFERENCES

1. B. Yan, G. Yue, X. Xu, J. Yang, and S. Guha, *Phys. Status Solidi* **207**, 671 (2010).
2. C.-H. Lee, A. Sazonov, and A. Nathan, *Appl. Phys. Lett.* **86**, 222106 (2005).
3. L. R. Wienkes, C. Blackwell, and J. Kakalios, *Appl. Phys. Lett.* **100**, 072105 (2012).
4. Y. Adjallah, C. Anderson, U. Kortshagen, and J. Kakalios, *J Appl Phys* **107**, 043704 (2010).
5. L. Mangolini, E. Thimsen, and U. Kortshagen, *Nano Lett.* **5**, 655 (2005).
6. J. H. Parker, D. W. Feldman, and M. Ashkin, *Phys. Rev.* **155**, 712 (1967).
7. D. Bermejo and M. Cardona, *J. Non-Cryst. Solids* **32**, 405 (1979).
8. P. D. Persans, A. F. Ruppert, B. Abeles, and T. Tiedje, *Phys. Rev. B* **32**, 5558 (1985).
9. A. G. Zabrodskii and I. S. Shlimak, *Sov. Phys. Semicond.* **9**, 391 (1975).
10. T. D. Moustakas, *J. Electron. Mater.* **8**, 391 (1979).
11. K. L. Chopra and S. K. Bahl, *Thin Solid Films* **12**, 211 (1972).
12. H. M. Dyalsingh and J. Kakalios, *Phys. Rev. B* **54**, 7630 (1996).
13. R. Street, *J. Electron. Mater.* **22**, 39 (1993).
14. D. K. C. MacDonald, *Thermoelectricity: An Introduction to the Principles* (Courier Dover Publications, 1962).
15. A. D. Stewart, D. I. Jones, and G. Willeke, *Philos. Mag. Part B* **48**, 333 (1983).

Mater. Res. Soc. Symp. Proc. Vol. 1536 © 2013 Materials Research Society
DOI: 10.1557/opl.2013.852

Low Temperature Annealing of Inkjet-Printed Silicon Thin-Films for Photovoltaic and Thermoelectric Devices

Etienne Drahi[1], Anshul Gupta[1], Sylvain Blayac[1], Sébastien Saunier[2], Laurent Lombez[3], Marie Jubault[3], Gilles Renou[3] and Patrick Benaben[1]

[1]Centre Microélectronique de Provence, Ecole Nationale Supérieure des Mines de Saint Etienne, 13541 Gardanne cedex, France
[2]Science des Matériaux et des Structure, Ecole Nationale Supérieure des Mines de Saint Etienne, 42023 Saint-Etienne cedex 2, France
[3]Institut de Recherche et Développement sur l'Energie Photovoltaïque (IRDEP), UMR 7174, EDF-CNRS-Chimie Paristech, 6 quai Watier, 78401 Chatou, France

ABSTRACT

Silicon nanoparticles-based inks were investigated in respect of their suitability for photovoltaic and thermoelectric applications. Nanoparticles with a diameter ranging between 20 to 150 nm were functionalized in order to avoid oxidation as well as having a good stability in suspension. After inkjet-printing and drying, they were annealed up to 1000 °C under nitrogen atmosphere by both rapid thermal and microwave annealing. The influence of the annealing treatment on the structural, electrical, optical and thermal properties was investigated by Raman, SEM, electrical and optical measurements. SEM and Raman demonstrate evolution of the microstructure at temperature as low as 600 °C. Optical, electrical and thermal properties depend strongly on the annealing temperature and tend to exhibit a modification of physical properties above 800 °C when the smallest nanoparticles begin to melt. The annealing method has been identified to be of primary importance on the layer microstructure and its thermal behavior.

INTRODUCTION

Amorphous (a-Si) and microcrystalline (μc-Si) silicon allowed the fabrication of thin films reducing the cost of fabrication of silicon solar cells. Nevertheless, it is possible to decrease the cost by avoiding vacuum-based process steps. In this objective, solution-based processes of silicon have a very strong potential being both low cost and large area compatible. More, inkjet printing process presents these advantages being a solution-based deposition process, but is also a non-contact, additive and maskless technology allowing patterning and working under atmospheric condition. Two types of silicon solutions are being investigated: liquid silicon-based on cyclopentasilane (CPS) precursor first demonstrated by Shimoda [1,2] and nanoparticles-based (NPs) suspensions [3,4]. Furthermore of its lower price compare to CPS, deposition of Si NPs suspension allows the tailoring of the properties by the advantage of modifying the size of the NPs, their doping level and sintering degree [3–5]. In the present study, Si NPs-based inks were inkjet-printed and sintered by two methods: a rapid thermal annealing (RTA) and microwave (μW) annealing under pure nitrogen (99.999%). The properties of the fabricated thin films were investigated by Raman, SEM, electrical and optical measurements with targeted applications such as photovoltaic and thermoelectric materials.

EXPERIMENT

The experimental work in this paper has been divided in three parts. Firstly, inkjet printing process and surface preparation are described for printability optimization. Secondly, the drying step has been applied in order to evaporate the ink dispersants and optimized to obtain homogeneous and continuous layers. These two steps were followed by an annealing one needed for recovering the functional properties by sintering the Si NPs. Two annealing methods have been studied: rapid thermal annealing (RTA) and microwave (μW) annealing. Thirdly, structural, thermal, electrical and optical properties of the annealed layers on quartz are characterized.

Source material and inkjet-printing process

A commercial suspension of Si nanoparticles (NPs) dispersed in ethylene glycol and water has been printed. The particles obtained by chemical synthesis are between 20 and 150 nm diameter and functionalized by Sodium Polymethacrylate (NaPMA) in order to avoid oxidation and guaranty their stability. The suspensions have a viscosity varying between ~13 cP at 23 °C and 5.5 cP at 50 °C and a superficial tension around 47 mN/m at 23 °C. A drop on demand (DoD) system (Dimatix printer DMP 2800) was used for printing of the Si NPs ink.

A pattern introduced in the printer program is turned into a data pulsed train, which is transmitted to a piezoelectric transducer. The piezoelectric changes his shape depending on the received electrical signal. This signal is formed by pulses, named waveforms, made of positive and negative voltage impulsions. The positive one dilates the piezoelectric allowing forming a drop of liquid from a nozzle. The negative one contracts the piezoelectric transducer which annihilates the fluid expansion giving birth to a drop. A specific waveform has been designed for the used ink.

Quartz lamellas with average roughness R_a=0.62 nm (AFM Veeco SP-II) were used as substrates. Acetone cleanings in ultrasonic bath has been applied as surface cleaning treatment before printing. A value of 47 mN/m for the surface energy has been measured by goniometric experiments. This cleaning step was observed as crucial for the formation of a homogeneous printed layer. Depending on the size of the printed drops on quartz substrates (~40 μm) spacing of 20 μm between two consecutive drop centers has been determined as being optimal distance. In order to increase the layer thickness, two passes printing has been done without intermediate drying or annealing.

Drying and annealing process

In order to recover the functional properties of the thin films, mandatory annealing has been performed using two methods: Rapid Thermal Annealing (RTA) and Microwave (μW) annealing. The RTA Jipelec FirstJet tool is equipped with 12 halogen lamps of 144 V and 1200 W. Concerning RTA, heating is performed by three mechanisms: infrared to ultraviolet radiation, heat convection and conduction [6]. A quartz wafer isolates the lamps and the chamber where the sample is placed.

The μW system and the temperature measurement setup have been described extensively in [7]. Heating is performed by direct absorption of the μW (2.45 GHz multimode) by the Si NPs but also by adding a SiC (home-made ring) susceptor in the cavity absorbing the μW and emitting infrareds. This causes the formation of thermocarriers in the Si NPs that enhance the absorption of μW [5].

Annealing in N_2 atmosphere at 600, 700, 800, 850, 900 and 1000 °C during 5 min have been applied with heating rate of 10 °C/s. A previous drying step is necessary in order to avoid condensation of ink, issuing from solvents, on the RTA quartz. A two steps drying has been observed to be optimal: a first step in vacuum (3-5 mbar) at room temperature during 10 min and a second step in N_2 atmosphere by RTA up to 200 °C during 1 min.

Characterization of annealed thin films

Structural, electrical and optical characterizations were realized. Structural evolutions were measured by Scanning Electron Microscopy (SEM) and Raman microscopy. SEM observations have been realized using a Carl Zeiss Ultra 55 SEM equipped with an Oxford Energy Dispersive X-ray Spectroscopy detector useful for chemical analysis. Raman measurements have been made using a Jobin Yvon LabRam 880 HR Raman microscope equipped with a 488 nm laser. By modifying the laser power, the thermal conductivity of the thin films was estimated.

Electrical measurements have been performed using a Keithley 4200 Semiconductor Parameter Analyzer. Large probes (round shape of about some hundreds micrometers in diameter) were used in order to limit degradation of the soft layers. Contacts have been taken directly at the corners of the printed pattern which results in a probing length of around 10 mm. Optical measurements in the range of 300 nm to 2000 nm were made using Perkin Elmer Lambda 900 UV/VIS optical spectrometer. A quartz substrate without nanoparticle layer was used as reference.

DISCUSSION

Structural analysis and thermal properties of annealed printed layers

Strong evolution of the layer microstructure (~1 μm thick for each sample) has been observed by SEM, this evolution depends on the annealing temperature, Figure 1 presents results obtained for RTA annealing under N_2 atmosphere and a dwell time of 5 min. The microstructure can therefore be tailored from a porous thin film with nanodomains to a denser thin film with microdomains. Cross-sections done on equivalent samples demonstrated formation of a denser layer between the substrate and the observed aggregates seen in Figure 1.

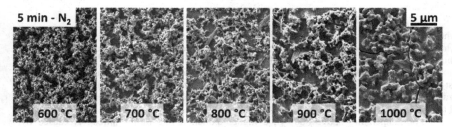

Figure 1 SEM pictures of the RTA samples under N_2 with 5 min dwell time.

These SEM pictures demonstrate that the morphology of the thin films can be tailored by applying different sintering steps. The high temperature annealed samples exhibit cracks in both the thin film and the substrate. They have been identified to be due to stress. Raman spectroscopy (at low laser power, <50 μW, in order to avoid induced heating of the layer) was used to estimate this stress by measuring the Raman shift $\Delta\omega$ from the following equation [8]:

$$\Delta\omega=\omega-\omega_0=-4.10^{-9}((\sigma_{xx}+\sigma_{yy})/2) \tag{1}$$

Where ω is the position of the Raman peak and ω_0 the reference position, σ_{xx} and σ_{yy} are the stress components (in Pa) in the x-y plane (thin film plane). Results are presented in Figure 2 (left) for both RTA and µW annealing.

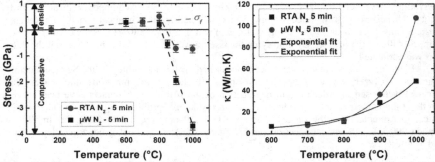

Figure 2 (left) Stress and (right) Thermal conductivity measurements on the printed and annealed Si NPs layers (RTA and µW) as a function of the annealing temperature.

Two phenomena can be observed: the thin film is in tension for temperature below 800 °C and in compression for temperature above 800 °C. While the compression can be explained by preferential oxidation at the grain boundaries therefore inducing compression on the Si grains [9], the tension (σ_f in Figure 2 left) results from the difference of thermal expansion coefficient between the quartz substrate and the Si thin film during cooling process. This can be estimated using the following equation:

$$\sigma_f = E_f.(\alpha_f - \alpha_s).\Delta T/(1-v_f) \tag{2}$$

where σ_f is the stress of the Si thin film (in the case of a bi-axial stress $\sigma_f = (\sigma_{xx} + \sigma_{yy})/2$), ΔT is the temperature gradient, α_s is the substrate thermal expansion coefficient (a value of 6.10^{-7} °C^{-1} was chosen for the quartz lamellas), E_f (163 GPa), v_f (0.23) and α_f ($2.6.10^{-6}$ °C^{-1}) are the Young modulus, Poisson coefficient and thermal expansion coefficient of the thin film [10]. In order to estimate the stress and compare it with Raman measurements, the values of c-Si, written in brackets, have been used. A good match between this estimation and measured values is obtained.

For temperature above 800 °C, a higher oxidation rate of the silicon is caused by elevation of the temperature and the melting of the first nanoparticles (≤20 nm) [5]. The µW annealed samples exhibit a marked lower stress. It can be either explained by a lower oxidation or coarser grains. In order to validate these hypotheses, thermal conductivity, that is a marker of the phonon dissipation and therefore of the grain size, has been measured by Raman microscopy in the following.

Under the probing laser, the Raman peak of silicon is shifted towards lower frequencies and broadens [11,12] due to heating. By using this property, Raman microscopy can be used to measure the effective temperature of a thin film [8]. The following linear relationship exists between the Raman shift and the temperature in c-Si:

$$\Delta\omega = T.(d\omega/dT) \qquad (3)$$

where $\Delta\omega$ is the Raman shift, T the effective temperature of the thin film and $d\omega/dT$ a factor different for each material (=-0.0242 cm^{-1}/K for c-Si [8]).

An estimation of the thermal conductivity of the Si NPs thin films can be found from the following equation [4] obtained by modifying the general equation developed by Nonnenmacher [13]. This new equation is only valid for thin films (thickness inferior to laser spot diameter)

$$\kappa = (d-1/2\alpha).P_{abs}/\Delta T \qquad (4)$$

where d is the thickness of the layer, α the absorption coefficient, P_{abs} the absorbed power of the laser beam and ΔT the temperature gradient (here between the temperature of the thin film and the room temperature).

The estimated thermal conductivities of the annealed layers are presented in Figure 2 (right). Values from some unities up to >100 W/m.K can be obtained. An exponential growth is observed for both type of annealing but µW annealed samples exhibit a much faster growth. This could also be attributed to coarser grains that would also explain the lower compressive stress.

Electrical and optical properties

I-V measurements have been perfomed on the annealed samples and exhibit an ohmic behavior. Therefore, equivalent electrical resistance of the layer can be plotted as a function of the annealing temperature and type (Figure 3 left).

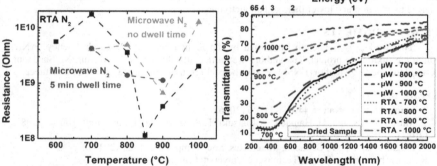

Figure 3 (left) Equivalent electrical resistances under dark (extracted from *I-V* curves ~1 cm of probing length) of Si NPs thin films as a function of the annealing temperature for RTA and µW annealing under N$_2$. (right) Transmittance of the annealed samples with 5 min dwell time

Decreasing of the electrical resistance (under dark conditions) is observed by increasing the annealing temperature down to a minimum resistance: this phenomenon is due to a densification and/or grain coarsening in the layer. Thereafter, the electrical resistance increases because of oxidation and cracking of the layer. The lower resistance is obtained for RTA under N$_2$ atmosphere. The highest electrical resistance of µW annealed samples can be explained by both a contamination in the furnace cavity (due to the crucible where sample were placed) and grain coarsening which induces a weaker densification of the layer, thus a difficult percolation of the carriers. No difference in electrical resistance is observed under light because of high recombination and low lateral conductivity of the thin films.

Concerning the optical measurements, the transmittance spectra show a strong increase as a function of the annealing temperature (Figure 3 right). This is attributed to the crystallization and densification of the material as well as cracking and possible holes (voids) throughout the thin film. Furthermore, increased inclusion of impurities such as oxygen could also provoke a widening of the optical gap. Higher transmittances are obtained by μW annealing compare to RTA. This phenomenon can also be explained by the grain coarsening that does not keep the grains nanosize, therefore decreasing the absorption coefficient. It also opens more holes throughout the layers than classical thermal annealing.

CONCLUSIONS

By applying a sintering step (rapid thermal or microwave annealing) the functional properties (electrical, optical and thermal) of inkjet-printed silicon nanoparticles thin films can be tailored from a nanosize porous material to a denser material with coarser grains. Strong changes of the structural properties appear around 600 °C. Electrical and optical properties, as well as stress in the layer, are observed at 800 °C where the melting of the smallest nanoparticles ($\leq$20 nm) and oxidation of silicon is observed. This material can be used either as an absorber layer for photovoltaic devices or as a thermoelectric material since thermal and electrical conductivity can be modified in a quite independent way. For photovoltaic applications the thickness of the layers should be increased and their purity improved to enhance both conductivity and recombination rate. For thermoelectric applications, RTA processing should be preferred thanks to much lower thermal conductivities.

ACKNOWLEDGMENTS

This work was financially supported by the French National Research Agency (ANR) through the Inxilicium project. The authors are particularly thankful to J. Mazuir, M. Saadaoui, T. Camilloni and D. Żymełka (ENSM-SE) for helpful discussions and experimental assistance.

REFERENCES

1. T. Shimoda, et al., Nature 440, 783–786 (2006).
2. T. Masuda, N. Sotani, H. Hamada, Y. Matsuki, T. Shimoda, Appl. Phys. Lett. 100, 253908 (2012).
3. E. Drahi, S. Blayac, P. Benaben, Mater. Res. Soc. Symp. Proc. 1321 (2011).
4. R.W. Lechner, PhD. Thesis, Technische Universität München, 2009.
5. E. Drahi, PhD. Thesis, Ecole Nationale Supérieure des Mines de Saint Etienne, 2013.
6. T.C. Kho, L.E. Black, K.R. McIntosh, in 24th European PVSEC, Hamburg, Germany, 2009.
7. D. Żymełka, S. Saunier, J. Molimard, D. Goeuriot, Adv. Eng. Mater. 13, 901–905 (2011).
8. I. De Wolf, J. Jiménez, J.-P. Landesman, C. Frigeri, P. Braun, E. Da Silva, E. Calvet, Raman and Luminescence Spectroscopy for Microelectronics, 1998.
9. M. Kawata, T. Katoda, J. Appl. Phys. 75, 7456–7459 (1994).
10. J.R. Greer, R.A. Street, J. Appl. Phys. 101, 103529 (2007).
11. R. Tsu, J.G. Hernandez, Appl. Phys. Lett. 41, 1016–1018 (1982).
12. M. Balkanski, R. f. Wallis, E. Haro, Phys. Rev. B 28, 1928–1934 (1983).
13. M. Nonnenmacher, H.K. Wickramasinghe, Appl. Phys. Lett. 61, 168–170 (1992).

Mater. Res. Soc. Symp. Proc. Vol. 1536 © 2013 Materials Research Society
DOI: 10.1557/opl.2013.890

Shape Evolution of Faceted Silicon Nanocrystals upon Thermal Annealing in an Oxide Matrix

Zhenyu Yang, Alexander R. Dobbie, and Jonathan G. C. Veinot
Department of Chemistry, University of Alberta, 11227 Saskatchewan Drive, Edmonton, Alberta, T6G 2G2, Canada

ABSTRACT

It is well established that controlled high-temperature annealing of hydrogen silsesquioxane leads to the formation of small spherical silicon nanocrystals (~3 nm). The present study outlines an investigation into the influence of annealing time and temperature. After prolonged annealing, crystal surfaces thermodynamically self-optimize to form a variety of faceted structures (*e.g.,* cubic, truncated trigonal and hexagonal structures).

INTRODUCTION

The synthesis of silicon nanoparticles/crystals has been a very active research area over the past 15-20 years in part because these materials are not accessible through the application of standard methods used to prepare traditional nanomaterials. In addition, their biocompatibility (compared to Cd-based quantum dots) make them particularly appealing. In this context, their unique properties make them suitable for nanoelectronic devices, *in-vivo* imaging, and other light-emitting applications [1-4].

Considerable effort has been aimed at controlling particle size and shape. By controlling the surface energy and electron transfer ability of quantum dots, their chemical and physical properties can be directly controlled [5, 6]. Morphological control has been widely studied and colloidal synthetic strategies are well-developed for metal and metal oxide nanocrystals, as well as II-VI and III-V quantum dots. Surfactants, temperature, and concentration allow the preferential growth of crystal faces by altering their relative thermodynamic stability. However, reports applying similar approaches to shape controlled synthesis of silicon nanomaterials are rare and even nonexistent because the strong directional bonding in Si precludes standard colloidal synthesis. The vast majority of the silicon nanoparticles synthesized by the decomposition of silane or other reduction-functionalization strategies are spherical or pseudospherical [7-10]. Synthesis of tetrahedral and cube shaped silicon nanostructures has been achieved using solution-based and nonthermal plasma methods, respectively [11-13]. However, the sizes of these nanostructures are relatively large and these procedures suffered from challenges such as flammable precursors and complicated infrastructure. Furthermore, our experimental understanding of the parameters that govern nanoparticle shape remains limited. Therefore, a new straightforward approach for making small Si-NCs of tailored shapes is appealing.

The Veinot research group has established a facile solid-state method that affords well-defined Si-NCs from hydrogen silsesquioxane (HSQ) as a precursor (Scheme 1) [14]. High temperature processing in a slightly reducing atmosphere causes HSQ to disproportionate and provides Si-NCs embedded in an SiO_2-like matrix. The size and crystallinity of these NCs may be tailored by defining the processing temperature [15]. Recently, we reported it is possible to exploit the relative thermodynamic stabilities of crystal faces to induce formation of silicon nanocubes inside the oxide matrix and that these cubes could be liberated via HF etching [16]. In the present study, we extend the synthesis of Si-NCs to include more complex morphologies (*e.g.*, hexagonal-shape) by tailoring the processing temperature and time. We also note the silicon oxide matrix influences Si-NC morphology and size: once the oxide matrix softens at higher temperatures, larger NCs (edge dimension >50 nm) of uncontrolled facetted structures form.

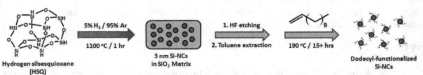

Scheme 1: Synthetic pathway from HSQ to dodecyl-functionalized silicon nanocrystals.

EXPERIMENTAL

Preparation of oxide-embedded silicon nanocrystals with various sizes and shapes. The solvent was removed from a HSQ stock solution (Purchased from Dow Corning as FOx-16) to yield white solid. 1g of the HSQ solid was placed in a quartz reaction boat and transferred to a Lindberg Blue tube furnace and heated from ambient to a peak processing temperature of 1100 °C at 18 °C/min in a slightly reducing atmosphere (*i.e.*, 5% H_2/95% Ar). The sample was maintained at the peak processing temperature for predetermined times to obtain particles of different dimensions (*e.g.*, 1 hour, d_{TEM} = 3 nm; 5 hours, d_{TEM} = 4 nm; 24 hours, d_{TEM} = 6 nm). Upon cooling to room temperature, the resulting amber solid was crushed using an agate mortar and pestle to break up larger pieces. More complete grinding was achieved using a Burrell Wrist Action Shaker upon shaking with high-purity silica beads for 5 hours. The resulting Si-NC/SiO_2-like composite powders were stored in standard glass vials.

The synthesis of larger particles (*i.e.*, d>6 nm) and Si-NCs with non-spherical shapes (*i.e.*, cubes, hexagonal and truncated trigonal prisms) required a second processing step at higher temperature. Following mortar and pestle grinding (*vide supra*), 0.5 g of Si-NC/SiO_2-like composite containing d_{TEM} = 3 nm Si-NCs were placed in a high temperature furnace (Sentro Tech Corp.) for further thermal processing in an argon atmosphere. Samples were heated to peak processing temperatures ranging from 1200 − 1400 °C at 10 °C/min and for processing times ranging between 1 and 72 hours. After cooling to room temperature, the resulting brown composites were ground using the identical procedures noted above.

Liberation of Si-NCs. Hydride-terminated Si-NCs were liberated from the protective SiO_2-like matrix upon etching with HF. Briefly, 0.25 g of ground/shaken composite was transferred to a polyethylene beaker equipped with a Teflon coated stir bar. 3 ml of water and 3 ml of ethanol was added under mechanical stirring followed by 3 ml of 49 % HF solution. After 1 hour, the color of the etching suspension changed from dark brown to orange/yellow. The liberated, hydride-terminated Si-NCs were isolated by extracting into *ca.* 30 ml (i.e., 3×10 ml) of toluene. The toluene solution was transferred to test tubes and used immediately for thermal hydrosilylation (*vide infra*).

Thermal hydrosilylation of Si-NCs. Following centrifugation at 3000 rpm, the toluene was decanted from the hydride-terminated Si-NCs and *ca.* 30 ml dodecene was added. The reaction mixture was transferred to a 100 mL Schlenk flask equipped with Teflon coated stir bar and the flask was evacuated and refilled with argon 3× to minimize the presence of air in the reaction solution. The reaction mixture was left open to an Ar filled manifold at atmospheric pressure, heated to 190 °C and was left stirring for a minimum of 15 hours.

Following thermal hydrosilylation, equal volumes (*i.e.*, ca. 7.5 ml) of the orange/yellow solution were placed in 4 centrifuge tubes. Approximately 35 ml of a 1:1 methanol:ethanol mixture was added to each. This procedure resulted in the formation of a cloudy yellow suspension. The precipitate was isolated by centrifugation in a high-speed centrifuge at 14000 rpm for 0.5 hour. The supernatant was decanted and the particles were redispersed in a minimum amount of toluene and re-precipitated by addition of 35 ml of 1:1 methanol:ethanol The dissolution/precipitation/centrifugation procedure was repeated twice. Finally, the purified functionalized Si-NCs were redispersed in toluene, filtered through a 0.45 μm PTFE syringe filter and stored in vials for further use.

Material characterization. Transmission electron microscopy (TEM) was performed using a JEOL-2010 (LaB$_6$ filament) electron microscope with an accelerating voltage of 200 kV. High resolution TEM (HRTEM) imaging was performed on JEOL-2200FS TEM instrument with an accelerating voltage of 200 kV. TEM samples of Si-NCs were drop-coated onto a holey carbon coated copper grid and the solvent was removed under vacuum. TEM and HRTEM images were processed using ImageJ and Gatan Digital micrograph software, respectively.

DISCUSSION

As a result of the present investigation, we have determined that Si-NC size and morphology obtained from thermal processing of HSQ and liberated from the resulting oxide matrix (Scheme 1) are dependent upon processing time and temperature. Based upon our observations we propose two complementary processes influence particle size/shape evolution: 1. Ostwald ripening leads to the formation of larger nanocrystals at the expense of smaller ones. 2. Diffusion and reorganization of silicon atoms to minimize the surface energy yield faceted structures. We also have found that higher temperature (*i.e.*, 1400 °C) annealing produces large faceted Si-NCs form, that presumable result of the combined influences of oxide matrix softening and melting of the silicon nanodomains which results in ready diffusion of Si atoms.

Influence of processing time and temperature

Table 1 summarizes how of Si-NC size and shape (i.e., frequency of cube formation) on thermal processing time and temperature. In general, particle size increases with longer processing time. For example, Si-NCs with d$_{TEM}$ = 3 nm and d$_{TEM}$ = 6.5 nm are obtained after 1-hour and 24-hour annealing at 1100 °C, respectively. The spherical/pseudospherical morphology of Si-NCs dominated even after 24-hour processing at 1100 °C. These observations are consistent with Ostwald ripening processes dominating particle evolution.

Raising the processing temperature to 1200 °C yielded faceted NCs when samples were processed for 24 hours (Figure 1a). The formation of nanocubes was noted after prolonged annealing (*i.e.*, 1200 °C for 72 hours, Figure 1b). These observations suggest the relative stability of Si crystal planes/faces are influencing particle shape. When the processing temperature was increased to 1300 °C surface self-optimization became more obvious. Processing at 1300 °C for 1 hr yields larger NCs (*i.e.*, d$_{TEM}$ = 7.9 nm) and the pseudospherical shape was maintained. Faceted particles were obtained from the samples processed for 15 hours and finally higher quantities of nanocubes (70%) were observed after 20-hour annealing. A similar trend was noted

209

for samples after annealing at 1350 °C, however faceted cubic structures were observed after much shorter processing times. Faceted NCs begin appearing after 1-hour processing and only a maximum of 21% nanocubes could be achieved.

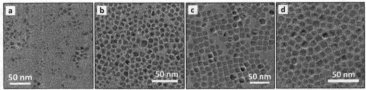

Figure 1: TEM images of dodecyl-functionalizaed faceted silicon nanocrystals formed from prolonged high temperature annealing: a) 1200 °C/24 hours, b) 1200 °C/72 hours, c) 1300 °C/20 hours, and d) 1350 °C/4 hours.

Table 1: Size and Percentage of nanocube structures in Si-NC samples.

Temperature (°C)	1100	1100	1200	1200	1200	1300	1300	1300
Time (hour)	1	24	1	24	72	1	10	20
Size (nm)	2.9	7.7	5.1	8.1	6.4	7.9	8.2	12.1
% Cubic	0	0	0	2	15	0	3	71

Temperature (°C)	1300	1300	1350	1350	1350	1350	1350	1350
Time (hour)	24	36	1	2	3	4	5	10
Size (nm)	11.4	8.7	8.0	7.5	7.6	7.4	6.6	7.0
% Cubic	34	7	3	11	16	21	8	7

Not surprisingly, prolonged annealing does not always yield more Si nanocubes. Fewer nanocubes were detected in samples prepared upon 24 and 36-hour processing at 1300 °C (*i.e.*, 34% and 7%, respectively). A similar trend was noted for samples annealed at 1350 °C. We contend that prolonged annealing at higher temperature results in rearrangement of silicon to yield non-cubic faceted NCs – a process that may be impacted by selective melting of the Si nanodomains and minimization of surface energy. This aspect of Si-NC shape evolution is the subject of ongoing investigation in our laboratory.

Formation of large faceted Si-NCs

Silicon atoms are expected to diffuse and rearrange readily at 1400 °C because the oxide matrix and Si nanodomains NCs are expected to soften or even melt. (SiO_2 softening point ca. 1200 °C, Si mp = 1414 °C). As a result, substantial morphology evolution of Si-NCs is observed (Figure 2a–c). After only 1-hour annealing a notable population of Si nanocubes (24%) was detected suggesting the oxide matrix is likely still intact and is influencing particle shape. The average size (edge dimension ~ 10 nm) is larger than those formed at lower temperatures, but a comparatively narrow size distribution remains consistent with a softened matrix and easier Si diffusion. No cubic structures were observed upon processing for 24 hours and only larger faceted structures (edge dimensions >20 nm) including, cuboid, hexagonal- and truncated trigonal-shapes were detected (Figure 2d). These observations suggest substantial Si diffusion and that these structures may have similar surface energy. Clearly, controlling silicon nanoparticle morphology within an oxide matrix is complicated by the softening of the oxide matrix which is expected to influence Si-NC size/shape upon high temperature annealing [10]. This is the subject of ongoing investigation in our laboratory.

Despite the complexity of these processes and their influences on particle size and shape evolution, valuable information can be obtained from straightforward observation of the particles obtained from various reaction conditions. At comparatively low temperatures (*i.e.*, ≤1350 °C) the rigid structure of oxide matrix is maintained and Si-NC size does change dramatically with increased processing time. In contrast, when the integrity of the SiO$_2$-like matrix, that effectively separates nucleation from growth when syntheses are performed at lower temperatures, is somewhat compromised at ca. 1400 °C comparatively rapid diffusion of Si atoms occurs and particle growth and shape evolution are promoted.

HRTEM images of Si particles obtained upon processing at 1400 °C are shown in Figure 2e-i. Larger nanocubes are formed at 1400 °C and they tend to grow in the [111] direction, consistent with the samples from 1300 °C reaction [16]. In addition, hexagonal and truncated-trigonal NCs with {111} bases and facets were found, consistent with the expected tetrahedral symmetry. The growth direction appears to be dominated by the thermodynamic stability of the {111}. Compared with previously reported large faceted Si-NCs (edge dimension >100 nm) [11, 12, 17], it is possible that these relatively small faceted particles may finally grow larger and finally similar morphologies if prolonged annealing is applied.

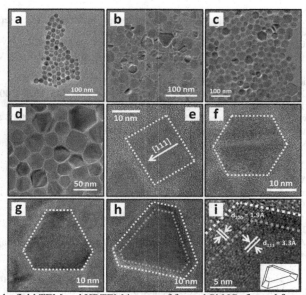

Figure 2: Bright-field TEM and HRTEM images of faceted Si-NCs formed from 1400 °C after a) 1-hour, b) 24-hour, and c, d) 48-hour annealing. Several types of faceted structures are shown: e) cuboid, f) hexagon, and g, h) truncated trigonal platelets. i) zoom-in image of h) showing two sets of fringes Inset: hypothetical structure of truncated trigonal platelets.

CONCLUSIONS

The present study demonstrates the formation of a variety of faceted Si-NCs upon high temperature annealing of oxide embedded Si-NCs obtained from thermal processing of HSQ. Particle evolution under high-temperature thermal annealing is complex. Prolonged annealing induces diffusion of surface atoms of spherical Si-NC to self-optimize and yield faceted structures. The optimization of processing time and temperature can effectively form nanocubes in an oxide matrix. At higher temperatures, the silicon nanodomains melt and surface atom diffusion occurs more readily and produces complex nanostructures. The oxide matrix plays an important role on controlling NC sizes upon annealing, while its softening would induce the formation of larger faceted structure with relative lower surface energy.

ACKNOWLEDGMENTS

We acknowledge the funding from the Natural Sciences and Engineering Research Council of Canada (NSERC), Canada Foundation for Innovation (CFI), Alberta Science and Research Investment Program (ASRIP), and the University of Alberta Department of Chemistry. G. Popowich and K. Cui are thanked for TEM and HRTEM analysis. All Veinot group members are thanked for useful discussions.

REFERENCES

1. G.v. Maltzahn, J.H. Park, K.Y. Lin, N. Singh, C. Schwöppe, R. Mesters, W.E. Berdel, E. Ruoslahti, M.J. Sailor, and S.N. Bhatia, *Nat. Mater.* **10**, 545 (2011).
2. Z.C. Holman, C. Liu, and U. Kortshagen, *Nano Lett.* **10**, 2661 (2010).
3. F. Erogbogbo, K.T. Yong, I. Roy, G.X. Xu, P.N. Prasad, and M.T. Swihart, *ACS Nano* **2**, 873 (2008).
4. X. Chen, S. Shen, L. Guo, and S.S. Mao, *Chem. Rev.* **110**, 6503 (2010).
5. C. Burda, X. Chen, R. Narayanan, and M.A. El-Sayed, *Chem. Rev.* **105**, 1025 (2005).
6. Z. Zhuang, Q. Peng, and Y. Li, *Chem. Soc. Rev.* **40**, 5492 (2011).
7. K.A. Pettigrew, Q. Liu, P.P. Power, and S.M. Kauzlarich, *Chem. Mater.* **15**, 4005 (2003).
8. X. Li, Y. He, and M.T. Swihart, *Langmuir* **20**, 4720 (2004).
9. J.H. Warner, A. Hoshino, K. Yamamoto, and R.D. Tilley, *Angew. Chem., Int. Ed.* **44**, 4550 (2005).
10. C.M. Hessel, D. Reid, M.G. Panthani, M.R. Rasch, B.W. Goodfellow, J. Wei, H. Fujii, V. Akhavan, and B.A. Korgel, *Chem. Mater.* **24**, 393 (2012).
11. R.K. Baldwin, K.A. Pettigrew, J.C. Garno, P.P. Power, and S.M. Kauzlarich, *J. Am. Chem. Soc.* **124**, 1150 (2002).
12. C.A. Barrett, C. Dickinson, S. Ahmed, T. Hantschel, K. Arstila, K.M. Ryan, *Nanotechnology* **20**, 275605 (2009).
13. A. Bapat, C. Anderson, C.R. Perrey, C.B. Carter, S.A. Campbell, and U. Kortshagen, *Plasma Phys. Control. Fusion* **46**, B97 (2004).
14. C.M. Hessel, E.J. Henderson, and J.G.C. Veinot, *Chem. Mater.* **18**, 6139 (2006).
15. C.M. Hessel, E.J. Henderson, and J.G.C. Veinot, *J. Phys. Chem. C* **111**, 6956 (2007).
16. Z. Yang, A.R. Dobbie, K, Cui, and J.G.C. Veinot, *J. Am. Chem. Soc.* **134**, 13958 (2012).
17. J.R. Heath, *Science* **258**, 1131 (1992).

Mater. Res. Soc. Symp. Proc. Vol. 1536 © 2013 Materials Research Society
DOI: 10.1557/opl.2013.755

Crystallization Kinetics of Plasma-Produced Amorphous Silicon Nanoparticles

Thomas Lopez[1], Lorenzo Mangolini[1,2]

[1]Mechanical Engineering Department, University of California, Riverside

[2]Materials Science and Engineering Program, University of California, Riverside

ABSTRACT

The use of a continuous flow non-thermal plasma reactor for the formation of silicon nanoparticles has attracted great interest because of the advantageous properties of the process [1]. Despite the short residence time in the plasma (around 10 milliseconds), a significant fraction of the precursor, silane, is converted and collected in the form of nanopowder. The structure of the produced powder can be tuned between amorphous and crystalline by adjusting the power of the radio-frequency excitation source, with higher power leading to the formation of crystalline particles. Numerical modeling suggests that higher excitation power results in a higher plasma density, which in turn increases the nanoparticle heating rate due to the interaction between ions, free radicals and the nanopowder suspended in the plasma [2]. While the experimental evidence suggests that plasma heating may be responsible for the formation of crystalline powder, an understanding of the mechanism that leads to the crystallization of the powder while in the plasma is lacking. In this work, we present an experimental investigation on the crystallization kinetic of plasma-produced amorphous powder. Silicon nanoparticles are nucleated and grown using a non-thermal plasma reactor similar to the one described in [1], but operated at low power to give amorphous nanoparticles in a 3-10 nm size range. The particles are then extracted from the reactor using an orifice and aerodynamically dragged into a low pressure reactor placed in a tube furnace capable of reaching temperatures up to 1000°C. Raman and TEM have been used to monitor the crystalline fraction of the material as a function of the residence time and temperature. It is expected that for a residence time in the annealing region of approximately ~300 milliseconds, a temperature of at least 750 °C is needed to observe the onset of crystallization. A range of crystalline percentages can be observed from 750 °C to 830 °C. A discussion of particle growth and particle interaction, based on experimental evidence, will be presented with its relation to the overall effect on crystallization. Further data analysis allows extrapolating the crystallization rate for the case of this simple, purely thermal system. We conclude that thermal effects alone are not sufficient to explain the formation of crystalline powder in non-thermal plasma reactors.

INTRODUCTION

Non-thermal plasmas are recognized as an efficient tool for the production of silicon nanoparticles with interesting optoelectronic properties and with several important potential applications. Yet a clear understanding of the correlation between the process parameters and the particle structure (crystalline vs. amorphous) is lacking, and it is well known that the particle structure has a strong influence on the nanoparticle optoelectronic properties [4]. In general, at low plasma power amorphous powder is produced, while at higher power nanocrystalline silicon is obtained [5]. In particular, for the reactor described in [6], crystalline powder with particle size smaller than 5 nm is obtained within a residence time of only few milliseconds. This raises some questions about the mechanism leading to the formation of crystalline particles. While the crystallization kinetic of silicon has already been extensively studied for the case of amorphous silicon thin films [7], little is known for the case of silicon nanoparticles. The reduction in the melting point for nanoparticles is a well-known phenomenon [8]. As a result, the crystallization temperature is also depressed in the case of nanostructures compared to bulk material. The crystallization temperature of silicon nanoparticles was measured by *in-situ* TEM in [9]. For instance for the case of 5 nm particles, a crystallization temperature of 800 K was reported, and for the case of 10 nm particles the crystallization temperature increases to 1300 K. Still, this measurement gives information about a phase change that occurs over a time period of several seconds to minutes (the time required to perform TEM analysis), and thus does not give information about the kinetics of crystallization over a time period in the order of milliseconds. In this paper we obtain an empirical measurement of the crystallization rate of plasma-produced silicon nanoparticles. This data provides useful information for understanding the mechanism of crystallization of particles immersed in partially ionized gases.

EXPERIMENTAL

The experimental apparatus consists of two in-line reactors, a first reactor for the synthesis of the amorphous powder and a second reactor for its in-flight annealing. The first reactor consists of a non-thermal plasma continuous flow system based on the design described in [1]. The non-thermal plasma reactor consist of a Pyrex tube with a diameter of either 9.525 mm or 25.4 mm. Depending on the reactor diameter we use either a dual electrode that is placed 1.5 cm from the downstream vacuum fitting (smaller tube) or a single electrode with a distance of 15mm from the downstream fitting. The vacuum fittings are Swagelok Ultra-Torr fitting. The electrical input power, of 10 watts, is supplied by a 13.56 MHz power supply, which is coupled with a T-type matching network to minimize reflected power. An orifice is placed between the first stage of the apparatus (powder production) and the second stage (powder annealing). The orifice allows for particles created in the non-thermal plasma to be aerodynamically pulled to the second reactor via a pressure gradient. The second reactor consists of a 60 cm Pyrex tube with a diameter of 2.54 cm, also connected with Ultra-Torr vacuum fitting. The second reactor is placed in a Lindberg tube furnace, which has a 40 cm heated length, and is capable of reaching temperatures of 1000 ^{0}C. All gasses are supplied to the system by compressed gas cylinders, with flow rates controlled by MKS Mass-Flow Controllers. For the samples discussed in this paper the following gases where used: Silane (1.37% in Argon) and ultra-high purity argon, 99.999%. The entire system was under vacuum via rotary pump with a base pressure < 10 mTorr. System pressure was controlled by an automatic butterfly valve that was placed downstream of the second reactor. Capacitance pressure gauges are used to monitor the system pressure both upstream and downstream of the orifice. The pressures in the first and second stage of the reactor are given in Table I. The total reactor volume is 3648 cm^3 to 3125 cm^3 depending on first stage reactor size, large or small, respectively. The typical leakage rate is < 1 mTorr per minute.

Table I: Production parameters

Sample	Tube diameter (mm)	Ar (sccm)	SiH4 (sccm)	Orifice (mm)	Pressure (Torr) Up/Down	Temperature Range ^{0}C
A	25.4	70	30	1.0	7.1/1.0	850-750
B	9.525	70	30	1.0	7.5/1.4	800-730
C	9.525	70	30	1.5	3.7/1.4	780-680

Two types of characterization were performed on the synthesized particles. First, particles were collected in flight on lacey carbon grids that were placed in the flow path of the particles at the exit of the second reactor, and TEM was performed on a Tecnai T12 120kV transmission electron microscope. Second, larger powder samples were collected in flight on fine stainless steel mesh that was placed at the exit of the second reactor. These sample collections were taken for longer periods of time producing 25 to 50 milligrams of powder in order to perform Raman Spectroscopy. Raman characterization was performed on Horiba LabRam, with a 532 nm laser.

The particles discussed in this paper were all made using the same parameters for gas flow and dilution. These conditions were a flow of 70 sccm of high-purity Argon and a flow of 30 sccm of 1.37% silane to Argon ratio. Table I shows the changes in tube diameter, orifice size, upstream and downstream pressure, and temperature ranges.

DISCUSSION

The crucial aspect in our work is the measurement of the crystalline fraction in the sample as a function of the tube furnace temperature in the second stage of the experimental apparatus. Crystallinity of the particles produced was tracked in two ways. First, TEM was performed to simultaneously monitor particle size and crystallinity, as seen in Figure 1. Figure 1-A shows a sample taken at room temperature which is amorphous and has a mean particle size distribution of 9 nm. Figure 1-B shows a sample taken at 600^{0}C that is also amorphous but has a slightly larger mean particle size of 13 nm. Figure 1-C shows a sample taken at 900 ^{0}C the sample is fully crystalline and has a larger mean particle size of 18 nm. It is clear that particles produced under these parameters show an increase in size as a function of temperature. This is a consequence of the inevitable agglomeration that occurs in the second stage. It is well known that particle agglomeration and coagulation is an exothermic process because of the reduction in specific surface area [10]. We have verified that this energy release term does not affect our crystalline fraction measurement by running a series of control experiments with different dilution with argon. In this way we control the particle concentration in the second stage and the

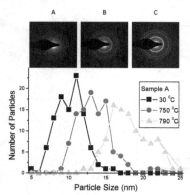

Figure 1: Size distribution & diffraction pattern

crystalline fraction measurement by running a series of control experiments with different dilution with argon. In this way we control the particle concentration in the second stage and the particle agglomeration rate. We have found no significant dependence between the particle concentration and the sample crystalline fraction, suggesting the crystallization rate is not affected by the particle agglomeration and coagulation. We have also made sure that new particles are not nucleated and grown in the second stage of the system. Precursor conversion efficiency was measured to ensure that all precursors were consumed in the non-thermal plasma. The efficiency was measured by collecting particles on fine mesh stainless steel filters, that were then massed and compared to the amount of incoming precursor gas. The efficiency was measured for multiple pressures and temperatures; the resulting conversion efficiency was 98% +/- 1%, for all samples. Based on these findings it is assumed, particle growth was due to particle interaction occurring at elevated temperatures in the second reactor.

Raman spectroscopy was used to track crystal percentage as a function of temperature. The samples were prepared by collecting powder downstream of the second reactor on stainless steel mesh filters. The samples were then transferred to a silicon wafer that was coated with 200 nm of copper to remove any signal from the silicon substrate. The samples were then pressed by hand, resulting in a Raman Spectroscopy measurement on semi-smooth surface. The Raman Spectra collected from these samples were processed using a three peak fitting procedure [11], as shown in figure 2. A departure from the most standard two peak procedure typically used for thin films is necessary for a system characterized by high specific surface area. The three peaks are centered at 480 cm^{-1} (amorphous silicon peak), at 512 to 520 cm^{-1} (crystalline silicon peak) and a 500 to 510 cm-1 peak, which is used to include the contributions from strained silicon bonds at the particle surface. We use Gaussian peak profiles, as seen in Figure 2, and we determine the crystalline percentage as a function of temperature using the following equation:

$$Crystal\ Percentag = \frac{(area_{510} + area_{520})}{(area_{480} + area_{510} + area_{520})}$$

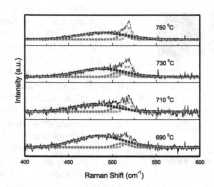

Figure 2: Raman spectra at different temperatures in the second stage, with Gaussian 3-peak fit

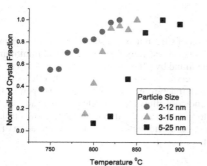

Figure 3: Normalized crystalline fraction versus temperature for different particle sizes

We renormalize the crystalline fractions determined by the Raman measurements by analyzing the TEM micrographs. Even for samples that appear to be composed solely of crystalline nanoparticles, the Raman measurement returns a crystalline fraction that is lower than 100%. This is due to the strong interaction between the laser beam used for the Raman measurement and the amorphous phase of silicon, compared to the crystalline phase. Figure 3 shows the normalized crystalline fraction as a function of temperature. The Arrhenius relationship of these two parameters allows for the extrapolation of activation energy for crystallization. These values are shown in Table II along with their corresponding particle sizes. Particle sizes listed correspond to particle size from the onset of crystallization until crystalline saturation occurs.

Table II: Crystallization Energy and Size Range

Sample	Energy (eV)	Size Range (nm)
A	4.73	15-25
B	2.06	10-17
C	0.93	7-12

Figure 4 is an Arrhenius plot of $Ln\left(\frac{Percent\ Crystalline}{Residence\ Time\ (\tau)}\right)$ versus the inverse temperature. The percent crystal value is taken from the data in figure #3 and divided by the residence time of

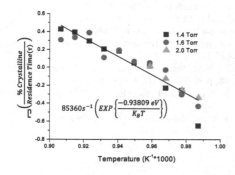

Figure 4: Crystallization rate of plasma-produced amorphous silicon nanoparticles

217

the tube furnace. A linear fit is then performed, with excellent agreement with the measured data. For figure 4 the particle size range is 3-8 nm. The particles were produced at three different downstream pressures of 1.4, 1.6, and 2.0 Torr. In the inset we show the crystallization rate as determined by the Arrhenius plot.

CONCLUSION:

The activation energy for crystallization of silicon nanoparticles with a size range of 3-12 nm was measured and found to be 0.93 eV. This value differs from a theoretical minimum value shown by [7], of 2.0 eV. This minimum value found by [7], was for the crystallization of amorphous atom clusters in thin films. The observations shown here justify that the behavior of crystallization kinetics is not just a function of temperature and time but is also dependent on geometry and size. Although plasma power increase allows for the release of surface terminating particles, little is still known as to how this would affect crystallization. Further research into these dependencies along with the characterization of surface termination of plasma produced nanoparticles is being investigated.

AKNOWLEDGEMENTS

Research supported by the U.S. Department of Energy, Office of Basic Energy Sciences, Division of Materials Sciences and Engineering under Award DE-SC0008934TDD

REFERENCES

1. Mangolini, L.; Thimsen, E.; Kortshagen, U., *High-yield plasma synthesis of luminescent silicon nanocrystals*. Nano Letters **2005**, *5* (4), 655-659.
2. Mangolini, L.; Kortshagen, U., *Selective nanoparticle heating: another form of nonequilibrium in dusty plasmas*. Physical review E **2009**, *79*, 026405 1-8.
3. Sriraman, S.; Agarwal, S.; Aydil, E. S.; Maroudas, D., *Mechanism of hydrogen-induced crystallization of amorphous silicon*. Nature **2002**, *418*, 62-65.
4. R. Anthony and U. Kortshagen, *Photoluminescence quantum yields of amorphous and crystalline silicon nanoparticles*. Physical Review B, **80**(11).(2009)
5. O. Yasar-Inceoglu, T. Lopez, E. Farshihagro, and L. Mangolini, *Silicon nanocrystal production through non-thermal plasma synthesis: a comparative study between silicon tetrachloride and silane precursors*. Nanotechnology, **23**(25): p. 255604.(2012)
6. L. Mangolini, E. Thimsen, and U. Kortshagen, *High-yield plasma synthesis of luminescent silicon nanocrystals*. Nano Letters, **5**(4): p. 655-659.(2005)
7. C. Spinella, S. Lombardo, and F. Priolo, *Crystal grain nucleation in amorphous silicon*. Journal of Applied Physics, **84**(10): p. 5383-5414.(1998)
8. A.N. Goldstein, C.M. Echer, and A.P. Alivisatos, *Melting in semiconductor nanocrystals*. Science, **256**(5062): p. 1425-1427.(1992)
9. M. Hirasawa, T. Orii, and T. Seto, *Size-dependent crystallization of Si nanoparticles*. Applied Physics Letters, **88**: p. 093119/1-093119/3.(2006)
10. K. E. J. Lehtinen and M. R. Zachariah, Physical Review B **63** (20), 205402/205401-205402/205407 (2001).
11. Qijin Cheng, *Crystal Growth & Design, Vol. 9, No. 6, 2009* **2865(1)**

AUTHOR INDEX

SUBJECT INDEX

absorption, 3
actuator, 79
amorphous, 17, 57, 73, 91, 113, 127,
    139, 147, 181

chemical vapor deposition (CVD)
    (deposition), 133
crystal growth, 207, 213

devices, 73, 79
dopant, 133

elastic properties, 147
electrical properties, 181
energy generation, 39

laser annealing, 45
luminescence, 105

microelectro-mechanical systems
    (MEMS), 155
microstructure, 175

nanostructure, 195, 207, 213

optical properties, 127, 139
optoelectronic, 39, 79, 85, 91

passivation, 119, 187
photoconductivity, 119
photovoltaic, 3, 17, 27, 51, 57, 63,
    105, 169
plasma-enhanced chemical vapor
    deposition (PECVD)
    (deposition), 27, 33, 51, 161,
    169, 195
porosity, 97

semiconducting, 85
sensor, 85, 91
Si, 17, 27, 33, 57, 73, 97, 113, 119,
    127, 147, 161, 169, 175, 181,
    187, 201, 207, 213
simulation, 187, 201, 207, 213
sintering, 97, 201
solution deposition, 51
sputtering, 139
stress/strain relationship, 155
structural, 113

texture, 45, 63
thermoelectricity, 195
thin film, 3, 33, 39, 105, 133, 155,
    161, 175, 201
transparent conductor, 45, 63

Printed in the United States
by Baker & Taylor Publisher Services